Green Energy and Technology

Climate change, environmental impact and the limited natural resources urge scientific research and novel technical solutions. The monograph series Green Energy and Technology serves as a publishing platform for scientific and technological approaches to "green"—i.e. environmentally friendly and sustainable—technologies. While a focus lies on energy and power supply, it also covers "green" solutions in industrial engineering and engineering design. Green Energy and Technology addresses researchers, advanced students, technical consultants as well as decision makers in industries and politics. Hence, the level of presentation spans from instructional to highly technical.

Indexed in Scopus.

Indexed in Ei Compendex.

Viktor Koval · Piotr Olczak
Editors

Circular Economy for Renewable Energy

Springer

Editors
Viktor Koval
Izmail State University of Humanities
Izmail, Ukraine

Piotr Olczak
Department of Minerals and Energy Market Research
Mineral and Energy Economy Research Institute of the Polish Academy of Sciences
Kraków, Poland

ISSN 1865-3529 ISSN 1865-3537 (electronic)
Green Energy and Technology
ISBN 978-3-031-30799-7 ISBN 978-3-031-30800-0 (eBook)
https://doi.org/10.1007/978-3-031-30800-0

This Springer imprint is published by the registered company Springer Nature Switzerland AG
The registered company address is: Gewerbestrasse 11, 6330 Cham, Switzerland

Preface

This book presents an exploration of the development of renewable energy sources as a circular resource in setting priorities for sustainable development, which is based on the environmental efficiency and energy balance of the circular economy model. Emphasis is placed on the importance of renewable energy by reducing consumers' costs, producing clean energy, and integrating it into the national power grid without compromising reliability or increasing energy security. After studying the circular economy model, readers are introduced to the problems of switching to alternative energy in the area of energy systems that cannot fully circulate all resources. The research results may allow stakeholders to adapt the sustainable renewable energy supply chain within the circular economy for the production of renewable energy equipment, which will have a synergistic effect on reducing carbon emissions.

The book serves as an interdisciplinary study linking the circular economy and renewable energy, satisfying the public interest in sustainable energy solutions from a social, political, economic, and technological perspective. All areas are considered to solve the problems of creating a sustainable energy system through the production of clean energy with integration into the national energy system without compromising reliability and improving energy security. This book also discusses a method for ensuring operational safety for renewable energy equipment, as well as models of renewable energy sources such as solar PV systems, battery energy storage systems, hydroelectric energy, bioenergy, and electric vehicles.

Editors want to express gratitude to all the contributors and reviewers who have aided in the publication of this book. Editors would also like to express our gratitude to Springer for their full support during the publishing process.

Izmail, Ukraine — Viktor Koval
Kraków, Poland — Piotr Olczak

Contents

An Approach to Ensure Operational Safety for Renewable Energy Equipment

Hanna Hrinchenko, Oleksandr Kupriyanov, Viktor Khomenko, Svitlana Khomenko, and Viktoriia Kniazieva

Abstract The task of nuclear energy at the present stage is to assess the technical condition and extend the service life of operating nuclear power units in order to ensure a smooth transition from a linear (traditional) economic model to a circular one. Nuclear energy is one of the cleanest and most powerful sources of renewable energy, but at the same time, there are questions about ensuring the safety of its use. To ensure the safety of the operation of energy complexes, the article proposes a method of technical diagnostics for pipeline systems based on the calculation of vibration resistance. Vibration oscillations occurring in the pipeline system of nuclear power plants during the movement of the coolant are a threat to violations of normal operating conditions, the occurrence of emergency shutdowns, and damage to the pipeline. The article proposes a vibration diagnostic method using the pipeline system for heating steam condensate from the high-pressure heater to the unit's deaerator as an example. When calculating the vibration load, the geometry of the pipeline system, pressure, velocity and thermophysical properties of the coolant that affect the amplitude of oscillations were taken into account. Based on the results of the calculation, recommendations were given to reduce the vibration load on the pipeline system, which makes it possible to extend the effective life of the power equipment in accordance with the approaches of the circular economy.

Keywords Renewable energy · Nuclear energy · Technical diagnostics · Safety · Pipeline systems · Vibrations

H. Hrinchenko (✉) · O. Kupriyanov · V. Khomenko · V. Kniazieva
Ukrainian Engineering Pedagogics Academy, Universiteyskaya St., 16, Kharkiv 61013, Ukraine
e-mail: hrinchenko@uipa.edu.ua

S. Khomenko
Berdyansk State Pedagogical University, Shmidt St., Berdyansk 71100, Ukraine

V. Koval and P. Olczak (eds.), *Circular Economy for Renewable Energy*, Green Energy and Technology, https://doi.org/10.1007/978-3-031-30800-0_1

1 Introduction

The depletion of non-renewable resources, which has become a global problem for mankind, is accompanied by serious environmental and social consequences, and the irrational use of resources and manufactured products leads to economic losses. In order to move to a more sustainable economic system, an approach to overcome the current linearity of the product life cycle through the alternative concept of a "circular economy" is increasingly discussed. The main idea behind this approach is to continue using materials and products instead of disposing of them as waste, thus ending their life cycle [1]. Economic growth in a circular economy can be achieved not by producing more resources (e.g., energy), but by extending the lives of existing capacities, namely energy equipment. An important task in forecasting and extending the service life is to ensure the safety of power equipment operation.

In Ukraine, the demand for renewable electricity is increasing every day, due to economic and social changes in the country [2]. Increasing needs and the impossibility of increasing energy capacity force to increase the efficiency of existing energy facilities. Ensuring the efficient use of renewable energy resources should be provided by reliable safety mechanisms. This is possible under the condition of constant technical diagnostics of the equipment of energy complexes.

The main technical parameters that influence the safe operation of a nuclear power plant (NPP) are vibration and erosion-corrosion wear of their pipeline systems as a result of transporting coolant (water, steam, and water), which leads to the accumulation of fatigue stresses. The accumulation of fatigue stresses, in turn, leads to the degradation of pipeline metal, reduces their vibration resistance, and, as a consequence, leads to the formation of fissures and breakage of fastenings, which leads to unscheduled repairs, downtime of the whole power unit, considerable material losses, and the emergence of the threat of accidents. As the reliability and operational safety of pipelines are connected with vibration, this characteristic can be used for the evaluation of the technical condition of pipeline systems as a whole and their elements.

Impact of significant vibrations on pipelines in conditions of stressed and deformed state of pipeline systems can lead to destruction of thermal insulation and hanger-support system, damage or jamming of valves, formation of fistulas and cracks on the pipes themselves, which leads to creation of emergency situations and unscheduled equipment shutdowns. Sometimes the fluctuations in pipelines are so significant that they prevent bringing the power unit to nominal capacity because of the threat of destruction or rupture of pipelines. To assess the loading of pipeline systems and structural elements due to vibration influences, it is necessary to determine the source of vibrations in order to reduce them and provide effective operation. The main causes of vibrations: the work of circulation pumps, transported coolant, the peculiarities of the design of pipeline systems. Determination of the cause of vibration and development of effective reduction mechanisms on the basis of technical diagnostics will ensure safe operation within the design timeframe and will contribute to a smooth transition to the over-design service life of all power equipment.

2 Literature Review

One of the priorities of economic development in the world is to ensure the efficient use of resources with minimal impact on the environment. The main impact on the environment when obtaining energy resources is due to carbon dioxide emissions.

The issues of analysis and scientific substantiation of nuclear energy as a renewable or "green" energy, taking into account the limitations of fossil resources, are becoming increasingly relevant. In 2022, the European Commission added nuclear energy to the taxonomy of "green" technologies as a type of economic activity considered sustainable in terms of environmental impact. However, many scientists still separate these concepts, although they agree that sustainable development is impossible without nuclear energy. For example, the paper [3] shows that the decrease in the share of nuclear energy in the global energy balance cannot be compensated by renewable energy sources and leads to an increase in the use of fossil fuels. To achieve sustainable non-carbon energy development, renewable and nuclear energy must be developed, where the risks of environmental consequences of energy production are minimal [3–5]. It was discovered that nuclear power plants (NPPs) play a significant role in the development of a non-carbon energy system, and that NPPs combine high environmental and economic efficiency with minimal risks and reliable operation safety.

The task of nuclear energy at the present stage is to assess the technical condition and extend the service life of operating nuclear power units to move away from the linear model and ensure a circular economy. Effective approaches are needed for the smooth transition of power supply operation to extending the service life while ensuring maximum safety of operation. Currently, much scientific and technical work is devoted to solving this problem.

For example, the authors of the studies [6, 7] propose to consider an approach to the formation of safety requirements on the basis of NPP design safety indicators. The design safety indicators of NPPs are those that can be determined during the design and controlled during NPP operation without requiring initial statistical data on the accident. The design features: power and fuel amount, determine the issues of quantitative determination of reactor plant safety indicators. The outlined approach to virtual accident assessment can be the basis for the development of appropriate normative documentation for constructive safety management of NPPs, excluding the statistical regularity of the accident, which indicates the need for technical diagnostics of power equipment and the development of quantitative safety indicators for individual elements and parameters.

In [8], it is proposed to improve the safety and efficiency of power plants by improving the regulatory support for cable products. Due to the need to modernize equipment at Ukrainian nuclear power plants, the replacement of cables, as an integral part of any system, significantly affects the safety of the entire energy complex. The paper systematizes the technical requirements of national regulatory documents on nuclear and radiation safety for cable products and defines the requirements for the selection of cables: fire safety, resistance to high temperatures, humidity, and pressure, resistance to ionizing radiation, seismic resistance, and electromagnetic compatibility. The proposed approach is important for the comprehensive safety assessment of nuclear power plants and the determination of the further effective service life of power equipment.

The issues of vibration load on pipeline systems from the point of view of determination and excavation are revealed in the work of the authors [9]. According to the relevant pipeline vibration measurement standards for nuclear power plant (NPP), a prototype pipeline vibration measurement system is designed and developed. The software based on signal processing technology and dynamic signal processing solution analysis is proposed. The collection of various vibration data from pipeline measurement points can provide effective data support for the subsequent construction of database, vibration evaluation, expert system and vibration reliability evaluation, which is very important to ensure the safety and reliability of the pipeline system. However, this study only proposes data collection without considering the causes of vibration loading. The issue of vibration and methods of its reduction are considered in the studies [10, 11]. The authors consider only the vibration caused by pumps as one of the most important factors affecting the safety and stability of nuclear power plant (NPP) operation, and it is also the most prominent vibration and noise transmission channel in the plant area [10]. During the study, it was determined that various support structures such as clamps, hangers, and pipe brackets are the primary means of transmitting pipeline vibration, and a method and scheme were proposed to effectively contain the low-frequency vibration of nuclear power plant pipeline systems. In this research, an electromagnetic semi-active vibration absorber is proposed that allows the electromagnetic stiffness to be adjusted by changing the excitation current, allowing the natural frequency to be changed online and the vibration damping effect to be improved. Assessment of the technical condition of power unit equipment requires a comprehensive approach. Such a comprehensive assessment is proposed by the authors to determine the quality of operation of pipeline systems, taking into account the peculiarities of their stress–strain state and seismic stability [12, 13]. Research methods: calculations using a mathematical model; comparison of calculation results, forecasting of limiting parameters of technical condition and monitoring of mechanical properties of material of the main pipeline; inspection, qualification, determination of the residual resource of buildings, structures, foundations and metal structures taking into account geotechnical and seismotectonic conditions. This research algorithm can be used as a basis for assessing the technical condition, namely its general model.

To ensure safety beyond the design term and the quality of operation of the power unit, it is necessary to create an effective mechanism for examining the technical condition, which would optimize its operation as much as possible, based on the favorable ratio of economic indicators and safety. In particular, such a mechanism will allow for a smooth transition to the operation of the power unit, without stopping it for a long time and the associated economic losses, in over-design periods [14, 15].

3 Research Methodology

The problem of increasing safe operation of NPP power units beyond the design basis is one of the most important and acute problems of nuclear power engineering. One of the most important elements of NPPs are pipelines carrying coolant, subjected to thermal, weight and vibration loads. Operational experience of NPP in Ukraine and abroad shows that pipeline vibrations reach considerable values and cause serious disturbances and directly influence NPP operation. One of the reasons of fatigue failure of pipelines is high vibration load of pipeline. The criterion of estimation of technical condition of pipelines is their calculation on vibration resistance which requires the data of vibration inspections.

Pipeline vibration inspections are followed by vibration strength calculations, and repair work is completed as planned or as problems arise. However, the planned replacement of equipment, including fittings, with more perfect and safe types may result in changes to the dynamic characteristics of the pipeline systems on which this work is carried out. The development of a method to assess the vibration resistance and vibration load of problematic pipeline sections and systems is topical. In this regard, a trim method for calculating the characteristic stresses of a rod system is considered, which makes it possible to assess the effect of vibrations on the system. If the amplitudes of alternating stresses are high enough, the elements of mechanical systems undergo fatigue damage accumulation, formation, and development, which leads to failure. Vibrations of pipelines at nuclear power plants during coolant transportation (water, steam, or water + steam) cause cracks and fissures in the metal of the pipelines, which can cause emergency situations, equipment failure, and significant material losses. In order to improve the safety of pipeline operation, it is proposed to calculate the vibration state of pipelines and, on its basis, develop recommendations for reducing vibrations to a safe level. The object of the inspection is a pipeline system for heating steam condensate from high-pressure heater PVD-6 to the deaerator of the unit after its reconstruction at the level rise in the section (level + 30.00; + 44.60; connection D-7) where vibrations with an amplitude of 5000 μm occurred. An axonometric diagram of the site is shown in Fig. 1.

Operating parameters of fluid in PVD-6: pressure $P = 15.5$ kgf/cm^2, temperature $t = 188.1$ °C; in deaerator D-7: pressure $P = 6$, kgf/cm^2, temperature $t = 164$ °C.

The pipeline material is steel 20. The mechanical and physical properties of the steel are given in Table 1.

Nominal mass flow rate $G = 520$ t/h. Prior to carrying out theoretical calculations, experimental measurements of vibration levels and frequency characteristics were carried out [13–18]. The measured data at minimum coolant level are presented in Table 2.

The pipeline under consideration, carrying condensate of heating steam, has the following properties of two-phase liquid flow. At temperature $t = 188.1$ °C and pressure in PVD-6 $P = 15.5$ kgf/cm^2 the liquid flow is below the saturation line which doesn't lead to intensive steam formation. When the liquid flow rises (as the pipeline is thermally insulated the process of liquid flow is adiabatic) at the level of

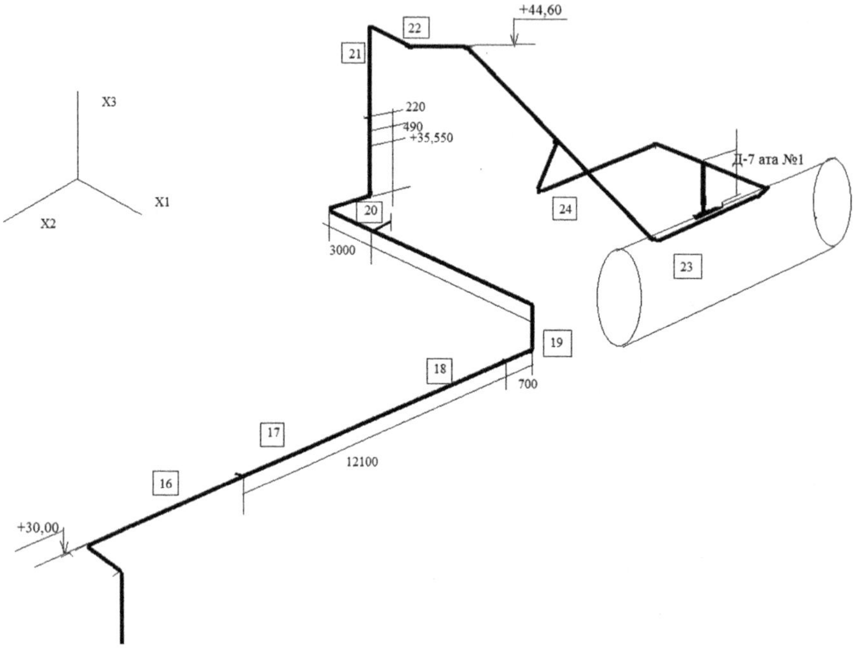

Fig. 1 Axonometric diagram of heating steam condensate pipeline from PVD-6 to deaerator of the unit at point + 30.00; + 44.60

Table 1 Mechanical and physical properties of steel 20

Parameter	Magnitude
Pipe geometric dimensions	Ø420 × 14
Modulus of elasticity	2.13 MPa
Density	7859
Poisson's ratio	0.28
Tensile strength	480 MPa
Yield strength	270 MPa

30 m the pressure will be P30 = 12.5 kgf/cm^2 the liquid flow is on the saturation line. Further rise of liquid flow level causes decrease of pressure P33 = 12.2 kgf/cm^2, P44 = 11.0 kgf/cm^2 and leads to intensive vaporization and growth of liquid flow speed. The measured data at vibration level and frequency characteristics at the support (point: + 40.00) are presented in Table 3.

Due to transient processes associated with local resistances (pipeline bends, welded joints, ovality), internal pressure changes occur in the system, with a pulsation frequency of f—0.5–6.4 Hz.

The velocity of the liquid flow is calculated from the nominal mass flow rate (139 kg/s) from the relation

Table 2 Vibration level and frequency characteristics at the supports and vertical section of the pipeline

Direction	Pipe support 17, (point + 30.00)		
	$V_{sliding}$, mm/s	A_{peak}, μm	*f*, Hz
X1	1.1	22	0.63–2.13
X2	2.7	120	0.63–2.13
X3	2.4	140	0.63–2.13
	Pipe support 19, (point + 31.00)		
X1	2.1	50	0.63–1.13
X2	6.2	400	0.63–1.13
X3	2.5	180	0.63–1.13
	Pipe support 20, (point + 32.80)		
X1	2.0	26	0.63–1.13
X2	8.5	190	0.63–1.13
X3	2.1	280	0.63–1.13
	Vertical section, (point + 35.50)		
X1	2.0	25	0.63–1.13
X2	8.5	1400	0.63–1.13
X3	6.8	1800	0.63–1.13

Table 3 Vibration level and frequency characteristics at the support (point: + 40.00)

Spring support 23, (point + 40.00). Vertical component[a]			
No.	$V_{sliding}$, mm/s	A_{peak}, μm	*f*, Hz
1	1.0	191	0.63
2	2.2	215	1.13
3	3.9	154	1.63
4	3.7	126	2.13
5	0.4	68	2.63
6	0.5	48	3.13
7	0.6	43	3.63
8	1.2	155	5.13
9	1.3	36	10.1

[a]The vibration levels were measured with a vibro-port VP-41

$$w = G_m / S \cdot \rho_{mix} \tag{1}$$

where Gm is the mass flow rate kg/s; S is the cross-sectional area of the pipeline; ρmix = ρw(1-φ) + ρπφ is the mixture density; ρw is the water density at operating temperature; φ is the steam content; ρπ is the steam density at operating temperature.

The densities of water and steam are determined from the tables of thermophysical properties of water and water vapour. The calculated data are shown in Table 4.

Table 4 Thermal properties of water and water vapour

№	w, m/s	ρ_w, kg/m^3	ρ_n, kg/m^3	φ, %	P_{mix}, kg/m^3
1	1.49	864.9	7.84	0.1	779.2
2	1.69	877.2	6.36	0.2	703.0
3	2.60	882.5	5.64	0.5	444.1
4	6.30	904.7	3.5	0.8	183.74

When the liquid flow moves (Fig. 2a), dynamic forces act on the pipeline, resulting in increased vibration (Fig. 2b).

In order to reduce the effect, additional stiffeners must be introduced or the piping routing must be changed, which is not possible on a working unit.

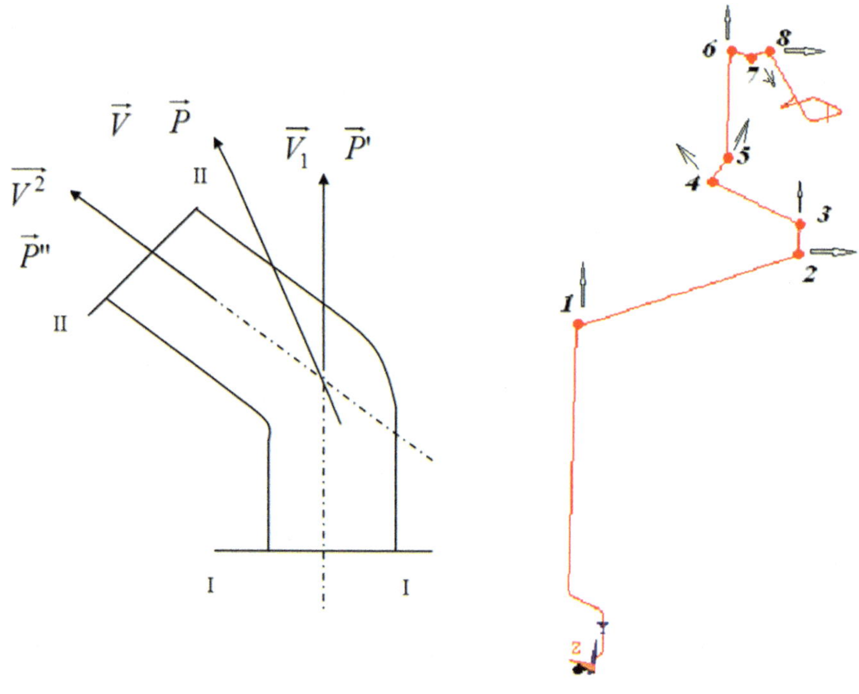

a *b*

Fig. 2 Forces acting on the pipeline (Force amplitude: 800 N on turns; external force frequencies: 0.5 and 1.25 Hz): a) At high-speed movement of the transported liquid in a bend; b) Axonometric diagram of the pipeline

4 Results and Discussion

The solution of the set task of calculating the vibration state of pipelines carrying heated steam condensate from PVD-6 to the deaerator of unit No. 2 was performed by the finite element method (FEM). A rod element with six degrees of freedom in a node was used for modeling the pipeline, and a rod element with tensile-compression properties with three degrees of freedom in each node was used for modeling the intermediate supports (Fig. 3) [19].

The eigenfrequencies waveforms obtained from the finite element model of the pipeline are shown in Fig. 4. Table 5 shows the eigenfrequencies of the pipeline.

Additional damping supports with rubber elements have been proposed. A calculation of the vibration condition before and after the retrofit was carried out.

The eigenfrequencies of the pipeline before (f1) and after (f2) the installation of the additional supports are presented in Table 6.

Figure 5 (points 1–8) show the results of calculation of forced vibrations before (figures with the index a) and after (figures with the index b) installation of additional supports. Displacements on different axes are highlighted with different colours. The vibration amplitudes have decreased by one or two orders of magnitude as a result of damping, indicating the need for rubber elements.

As a result of these tests, eigenfrequencies and forced oscillation amplitudes were determined, and additional damping supports were installed to reduce vibration levels.

Figure 6 shows the amplitude-frequency characteristics of the system after installation of the additional support. The data were obtained experimentally by measuring the vibro-port VP-41.

The methodology of experimental research included the determination of:

(1) pressure pulsations in the pipeline in different sections to identify the most loaded (in our case, it is a curved fragment of the pipeline);
(2) amplitude-frequency characteristics of the pipeline for comparison with the calculated data;
(3) amplitude-frequency characteristics of the pipeline after installation of additional supports.

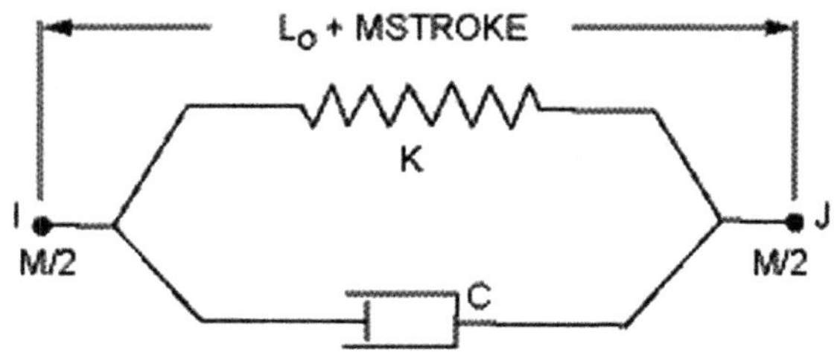

Fig. 3 Rod finite element

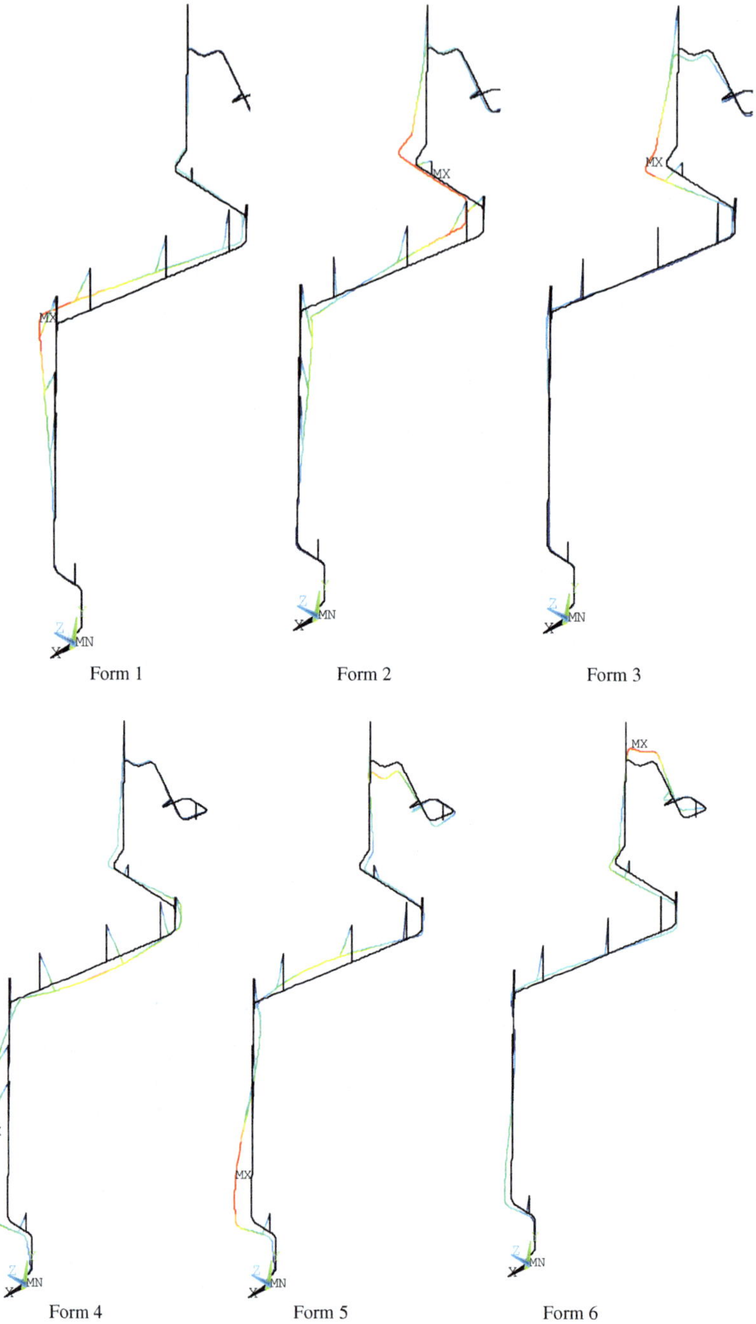

Fig. 4 Eigenfrequencies waveforms of the pipeline

Table 5 Eigenfrequencies of the pipeline

Frequency No	Frequency, Hz	Frequency No	Frequency, Hz
1	0.24	11	3.10
2	0.43	12	3.28
3	0.71	13	3.85
4	1.01	14	4.30
5	1.11	15	4.91
6	1.29	16	5.14
7	1.55	17	5.62
8	1.90	18	6.30
9	2.29	19	6.61
10	2.84	20	6.88

Table 6 Pipeline eigenfrequencies before (f1) and after (f2) retrofit

Frequency No.	$f1$, Hz	$f2$, Hz	Frequency No	$f1$, Hz	$f2$, Hz
1	0.24	0.68	11	3.04	4.65
2	0.43	0.85	12	3.55	4.79
3	0.56	1.09	13	3.86	5.16
4	0.77	1.32	14	4.30	5.78
5	1.04	1.61	15	4.82	6.02
6	1.17	2.31	16	5.04	6.84
7	1.41	2.47	17	5.25	7.35
8	1.67	3.07	18	5.69	7.47
9	2.28	3.48	19	6.45	8.72
10	2.34	3.86	20	6.84	9.07

Experimental studies confirmed the calculations in that the highest pressure pulsations occur in the curved section of the pipeline. The size of pressure pulsations in this section is 14 kPa, which leads to significant pipeline vibrations.

At the next stage, the vibration of the investigated pipeline section was controlled by determining the amplitude-frequency characteristics. For this purpose, piezoelectric accelerometers were used with a measuring range from 1 Hz to 10,000 Hz and a reduced error of 2.2% with an unevenness of the amplitude-frequency response of 3.8%.

Comparative results of experimental and calculated studies show a high degree of agreement, which amounted to 90%, which confirms the validity of the calculated data. The amplitude of vibration after the installation of additional supports decreased by an order of magnitude, while the excitation frequencies and natural frequencies of the pipeline differed significantly, which indicated the absence of resonance.

5 Conclusions

The method of emission diagnostics of pipeline systems of nuclear power plants is proposed in order to ensure the safety and efficiency the operation of the renewable energy facility. The methodology consists of building a finite element model and determining the eigenfrequencies and forced oscillations. As a result of the calculations, increased levels of vibration load during the transportation of heat flow were revealed, and it was proposed to install additional rubber supports to reduce them. According to the results of the technical diagnostics, after the installation of additional supports, the level of forced vibrations decreased in some areas by 10 times, and there was a general decrease in vibration load in the entire pipeline system. As a result of the calculation, the amplitudes of oscillations were obtained, which in some areas (point 6) were up to 2.5 m (along the axis OX).

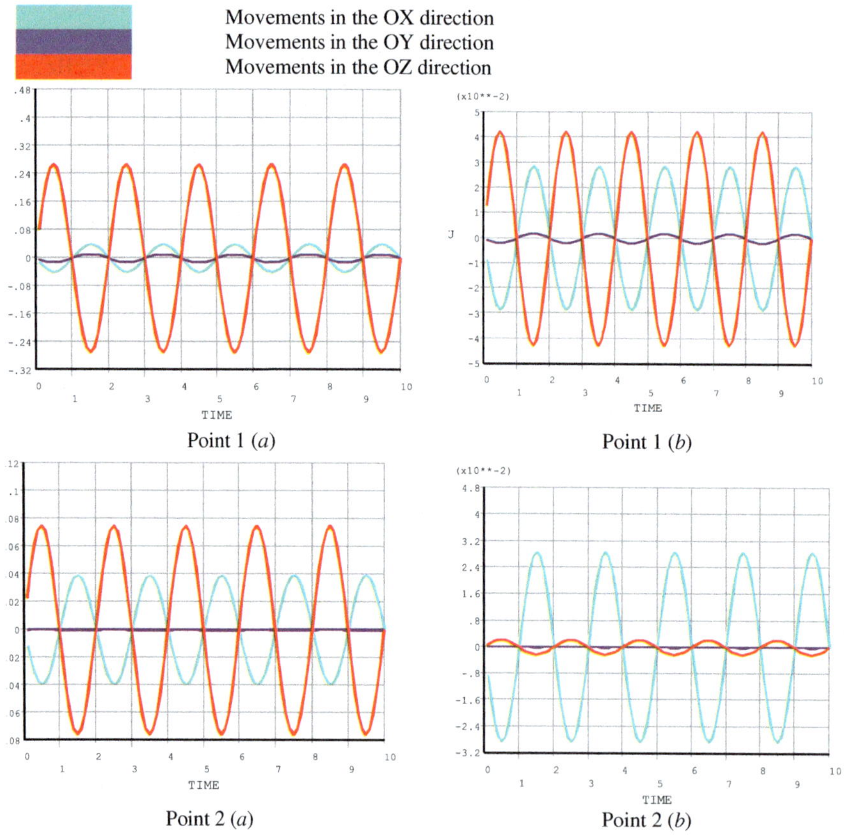

Fig. 5 Results of calculation of forced oscillations before (figures with index a) and after (figures with index b) in points 1–8

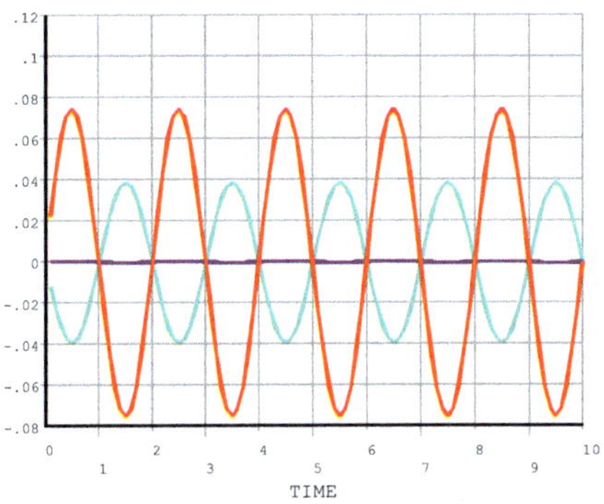

Point 3 (*a*)

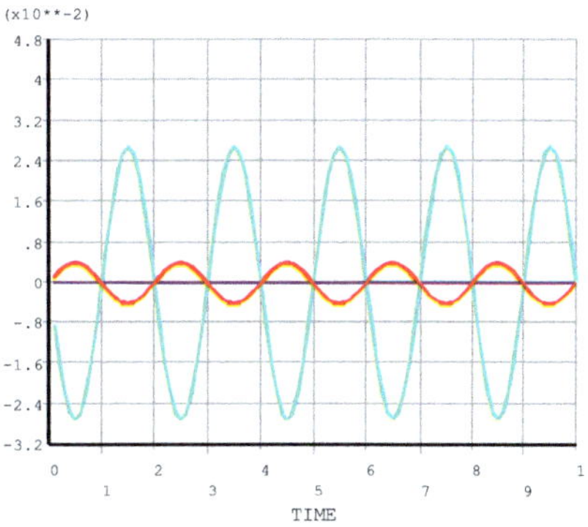

Point 3 (*b*)

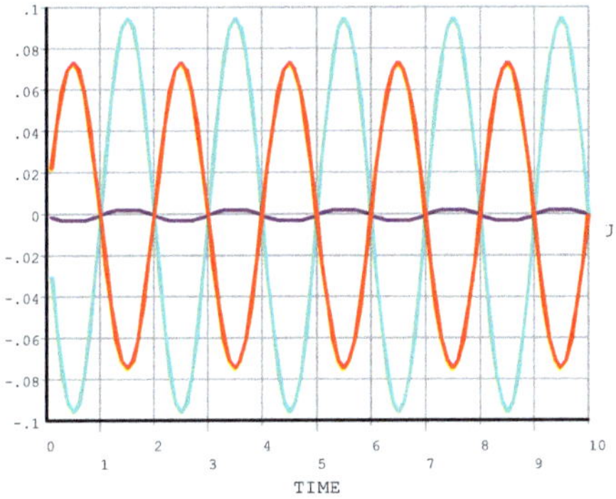

Point 4 (*a*)

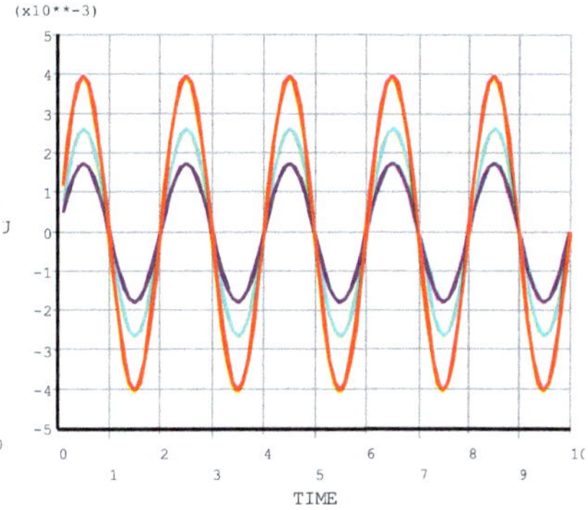

Point 4 (*b*)

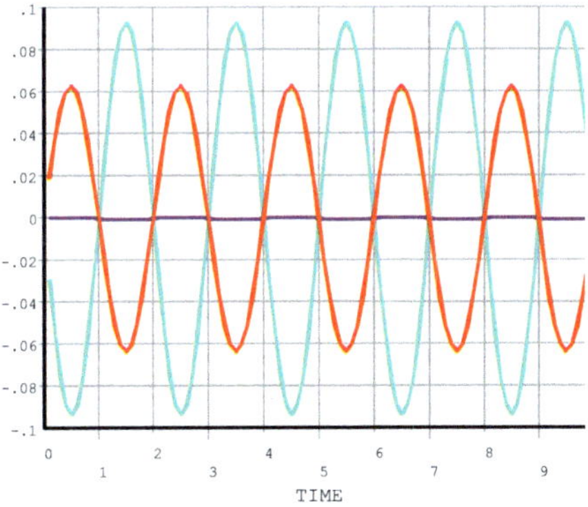

Point 5 (*a*)

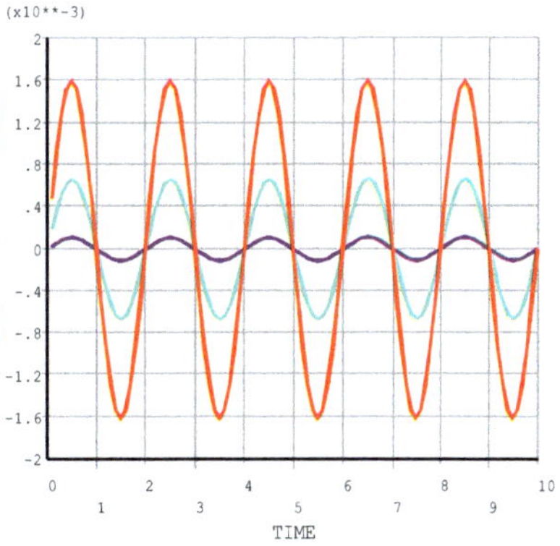

Point 5 (*b*)

Fig. 5 (continued)

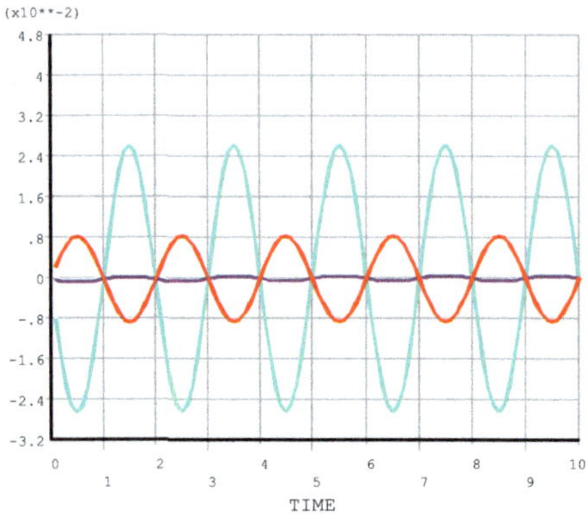

Point 6 (*a*)

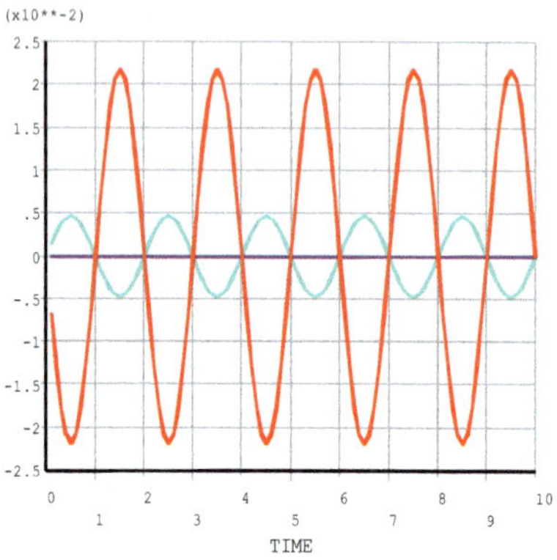

Point 6 (*b*)

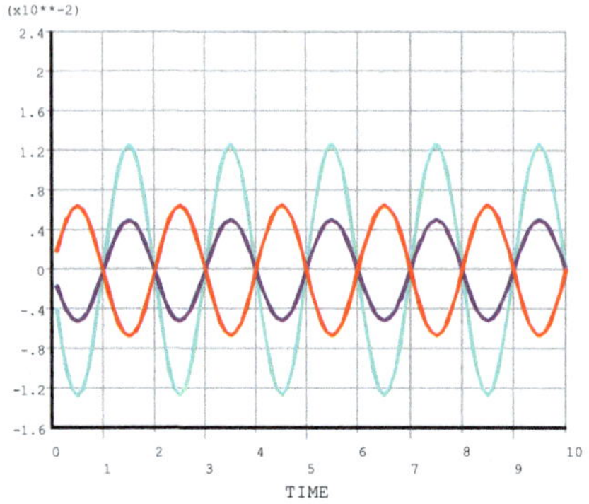

Point 7 (*a*)

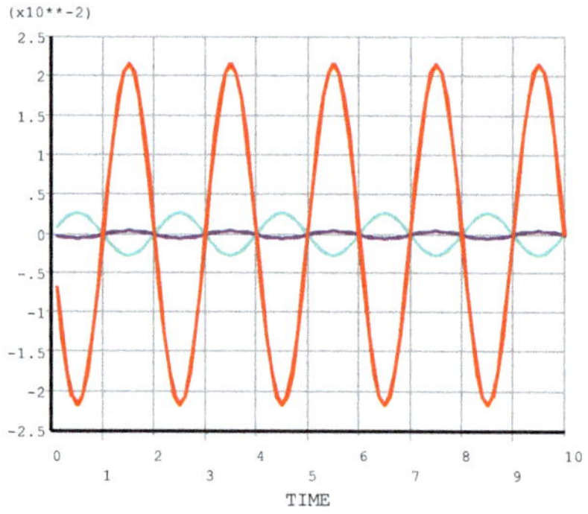

Point 7 (*b*)

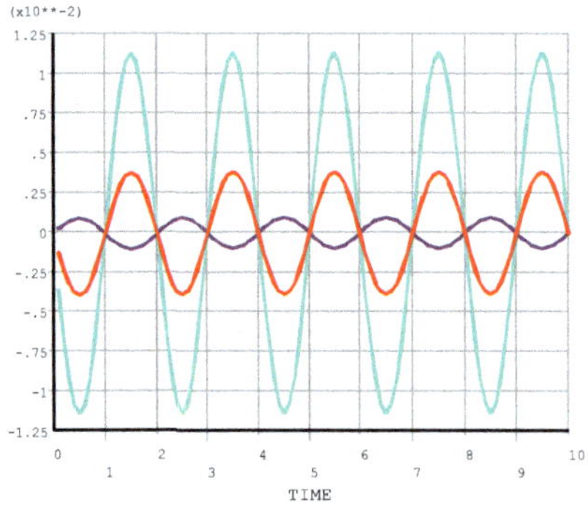

Point 8 (*a*)

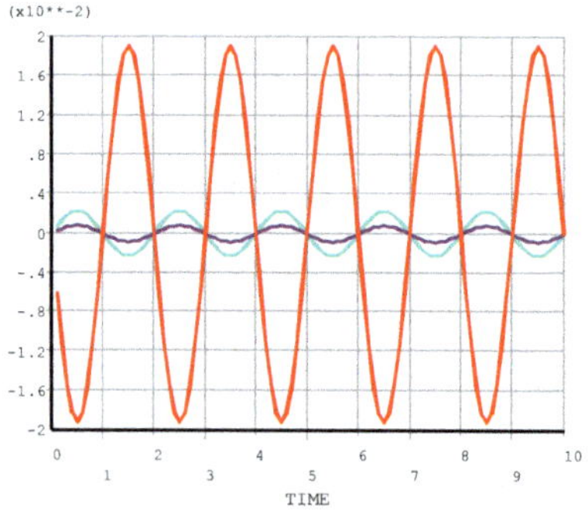

Point 8 (*b*)

Fig. 5 (continued)

After the installation of an additional rubber support, the amplitude of oscillations (point 6) decreased to a level of 0.5×10^{-2} m (along the axis OX). At the same time, there was a restoration of eigenfrequencies; for example, Form 1 changed the frequency from 0.24 to 0.68 Hz, which indicates the effectiveness of the use of rubber supports to avoid resonance. The proposed approach can be used as part of a comprehensive diagnostics of power equipment to ensure the safety of nuclear power plant operation both during normal operation and when operating beyond the design life.

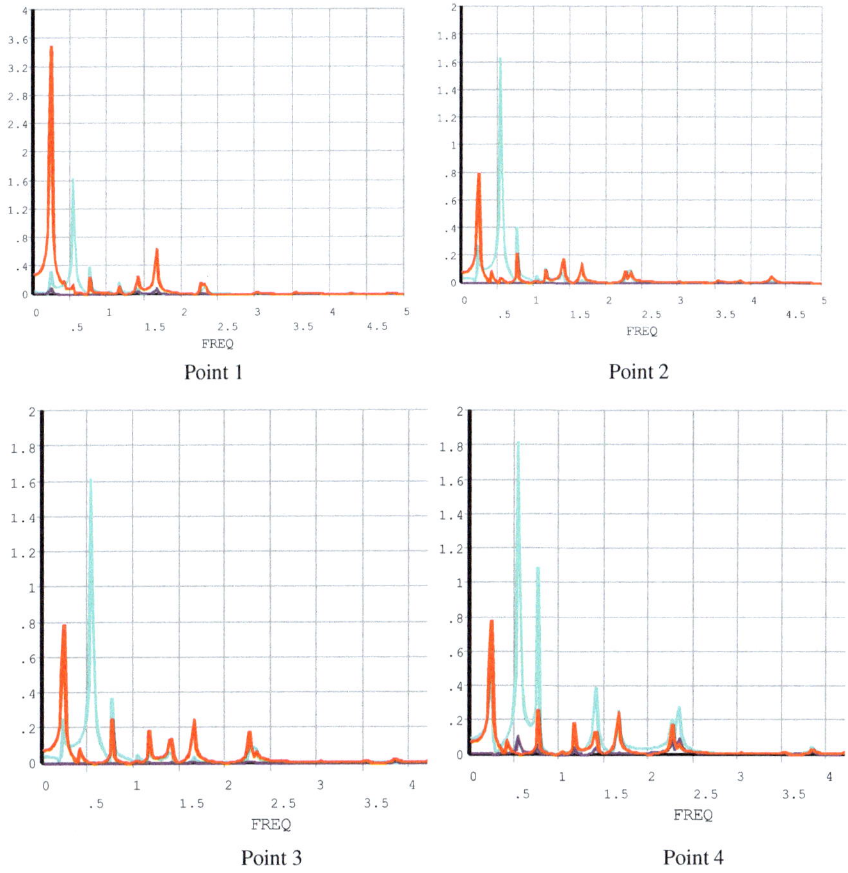

Fig. 6 Amplitude-frequency response of the system

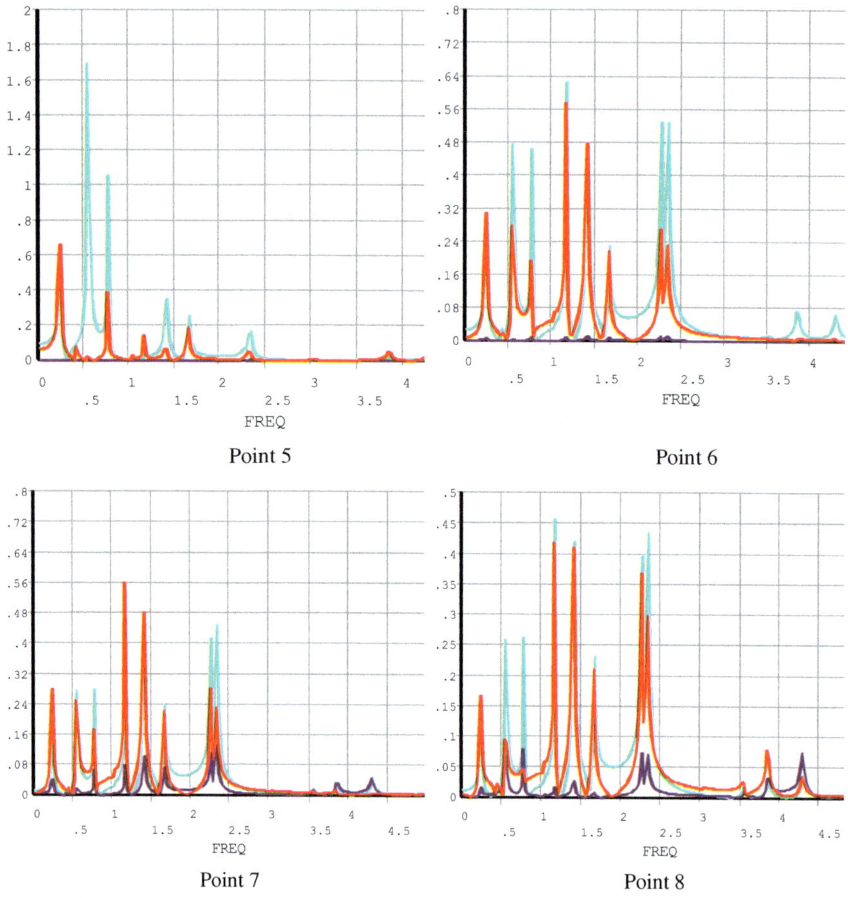

Fig. 6 (continued)

References

1. Geissdoerfer M, Savaget P, Bocken N, Hultink EJ (2017) The circular economy—a new sustainability paradigm? J Clean Prod 143:757–768. https://doi.org/10.1016/j.jclepro.2016.12.048
2. Ostapenko O, Olczak P, Koval V, Hren L, Matuszewska D, Postupna O (2022) Application of geoinformation systems for assessment of effective integration of renewable energy technologies in the energy sector of Ukraine. Appl Sci (Basel) 12:592. https://doi.org/10.3390/app12020592
3. Danish MSS, Gábor P (2022) Environmental and economic efficiency of nuclear projects. In: Clean energy investments for zero emission projects, pp 115–126. https://doi.org/10.1007/978-3-031-12958-2_10
4. Usman M, Radulescu M (2022) Examining the role of nuclear and renewable energy in reducing carbon footprint: does the role of technological innovation really create some difference? Sci Total Environ. 841:1–15. 156662. https://doi.org/10.1016/j.scitotenv.2022.156662

5. Baek J (2015) A panel cointegration analysis of CO_2 emissions, nuclear energy and income in major nuclear generating countries. Appl Energy 145:133–138. https://doi.org/10.1016/j.apenergy.2015.01.074
6. Borysenko VI, Klyuchnykov OO, Pampuro VI (2011) Background of NPP safety parameters. Safety problems of nuclear power plants and Chernobyl 5:6–12
7. Ginevičius R, Trishch R, Bilan Y, Lis M, Pencik J (2022) Assessment of the economic efficiency of energy development in the industrial sector of the European Union area countries. Energies 15(9):3322. https://doi.org/10.3390/en15093322
8. Iokst OA, Iokst VA, Ivashova LS, Yefimova KM, Levakin VV (2015) Basic requirements for cables of systems important to NPP safety. Nucl Radiat Saf 1:49–53. https://inis.iaea.org/search/search.aspx?orig_q=RN:46107931
9. Chao H, Lai J, Danping Yu, Xihua L, Sun L (2019) Research and development of vibration measurement system of pipeline in NPP. Nucl Power Eng. https://doi.org/10.13832/j.jnpe.2019
10. Zhiguo W, Jinlan G, Mengran L, Shaodan L, Lu D (2018) Optimal design and performance simulation of a novel semi-active vibration absorber for pipeline system of NPP. In: 26th International conference on nuclear engineering. London, England, 22–26 July 2018. V002T03A021. https://doi.org/10.1115/ICONE26-81741
11. Kun C, Jianrong Z, Yilin C, Jie Q, Changfan Z, Jian Q, Mingguang Z (2022) Vibration cause analysis and elimination of stagnant branch pipeline. Eng Fail Anal. 141. 106666. https://doi.org/10.1016/j.engfailanal.2022.106666
12. Hrinchenko H, Trisch R, Burdeina V, Chelysheva S (2019) Algorithm of technical diagnostics of the complicated damage to the continued resource of the circulation pipeline of the nuclear power plant. Prob Atomic Sci Technol, Section "Physics of Radiation Effect and Radiation Materials Science". 2(120):104–110. https://doi.org/10.46813/2019-120-104
13. Kiporenko G, Pakhalovich M, Khoroshilov O (2016) Evaluation of the technical camp of NPP piping systems for compliance with normative parameters. Control Syst, Navig Sound 4:146–152
14. Kiporenko G, Pakhalovich M (2016) Improvement of method of calculation resistance to brittle fracture pipelines South-Ukrainian NPP. Inf Process Syst 7:181–184. https://www.hups.mil.gov.ua/periodic-app/article/16946/eng
15. PM-T.0.08.139-04 (2004) Program for assessing the technical condition and reassignment of the resource of typical equipment of nuclear power plants. Turbogenerator
16. Cabinet Ministry of Ukraine: Comprehensive (Integrated) Safety Improvement Program for Ukrainian NPPs: approved by Cabinet Resolution No. 1270 of 07.12.2011 (2011). https://zakon.rada.gov.ua/laws/show/1270-2011-%D0%BF#Text
17. ISO 5725-2:2019: Accuracy (trueness and precision) of measurement methods and results—part 2: basic method for the determination of repeatability and reproducibility of a standard measurement method (2019)
18. Trishch R, Maletska O, Hrinchenko H, Artiukh S, Burdeina V, Antonenko N (2019) Development and validation of measurement techniques according to ISO/IEC 17025:2017. In: 2019 IEEE 8th International conference on advanced optoelectronics and lasers (CAOL), pp 1–6. https://doi.org/10.1109/CAOL46282.2019.9019539
19. Kupriyanov O (2019) Ansys simulation of the joint strength with the interference fit in the presence of the shape geometry error. Lecture Notes in Mechanical Engineering. Springer International Publishing, Cham, pp 72–78. https://doi.org/10.1007/978-3-319-93587-4_8

The Use of Electricity Storage on the Path to Prosumer Energy Self-sufficiency: Eastern Europe Case Study

Piotr Olczak, Viktor Koval, Viktoriia Yanovska, and Iryna Lomachynska

Abstract The increase in photovoltaic (PV) installations sets them apart from other sources of electricity in end-use industries due to increased capacity. The market for photovoltaic installations in Eastern Europe is rapidly developing due to both private investment and government support, as a result of which the share of solar energy has increased significantly and can reach the new potential of photovoltaic installations can reach 10 GW of average annual power. As part of home PV installations, the coefficient of energy self-consumption is about 30%. It can be increased by various initiatives, including energy storage. The higher the value of the auto-consumption coefficient, the lower the negative impact of the installation on the energy flows in the network. The article analyzes the impact of energy storage of different sizes on the auto-consumption rate. An algorithm of energy flows in the installation was prepared, taking into account energy storage. The analysis carried out showed that the use of an energy storage with a usable capacity of 120 kWh only slightly increases the possibility of increasing the amount of self-consumed energy (from 64 to 70%) compared to a storage with a capacity of 9 kWh.

Keywords Energy storage · Photovoltaic · Household · Self-consumption

P. Olczak (✉)
Mineral and Energy Economy Research Institute of the Polish Academy of Sciences, 7A Wybickiego Str., 31-261 Cracow, Poland
e-mail: olczak@min-pan.krakow.pl

V. Koval
Izmail State University of Humanities, 12 Repina Str., Izmail 68610, Ukraine

V. Yanovska
State University of Infrastructure and Technologies, Ivan Ohiienko Str. 19, Kyiv 03049, Ukraine

I. Lomachynska
Odesa National I.I. Mechnikov University, 2 Dvoryanska Str., Odesa 65082, Ukraine

V. Koval and P. Olczak (eds.), *Circular Economy for Renewable Energy*, Green Energy and Technology, https://doi.org/10.1007/978-3-031-30800-0_2

1 Introduction

1.1 Photovoltaic and Energy Storage

The Eastern European energy market has a significant option for the development of the renewable energy sector, which guarantees the stability of the energy system and uninterrupted energy supply to consumers. In Poland, in the years 2019–2022, over a million prosumer PV micro-installations with a total capacity of about 9 GWp were built (Fig. 1) [1]. Support programs (e.g. «My Electricity» [2], «Auction System» [3]) and a decrease in the cost of producing energy from photovoltaics [4] contributed to such an increase in the number of installations. In addition, the reason was the search by representatives of various economic sectors for the possibility of obtaining cheap energy [5], expressed by the LCOE indicator [6, 7]. In Ukraine, the total volume of commissioned facilities using renewable energy sources (RES) based on photovoltaic installations at the beginning of 2020 amounted to 2640 MW (73%) of the total RES installed on the territory of Ukraine [8].

In September 2022, the sum of electric power installed in PV exceeded 11 GWp, and the amount of energy produced by photovoltaic installations in several months of 2022 exceeded 1 TWh (Fig. 2.) [9]. In the coming years, the amount of energy supplied by PV should increase, however, one of the barriers to increasing the installed capacity is the possibility of transmitting energy during periods of high productivity of photovoltaic installations [10]. One of the ways to reduce excessive energy flows from prosumer installations is to increase the coefficient of energy self-consumption by, for example, using energy storage.

In the field of energy storage and PV micro-installations in Poland, the My Electricity 4.0 support program for households was introduced in 2022 [11].

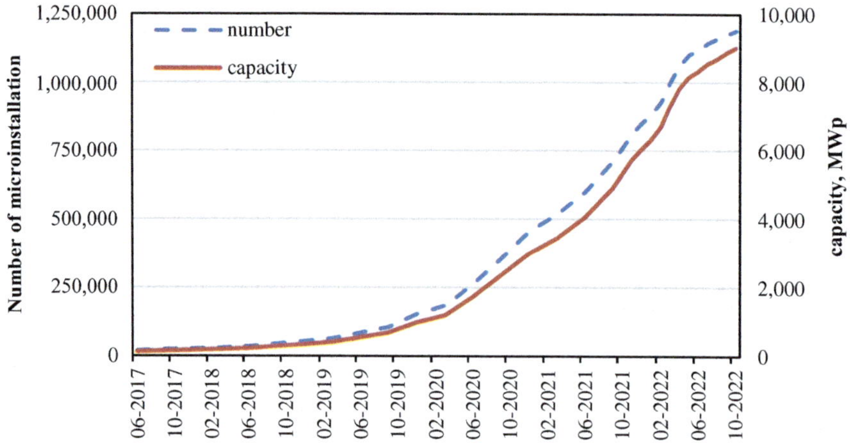

Fig. 1 Number and capacity of PV microinstallations in Poland [1]

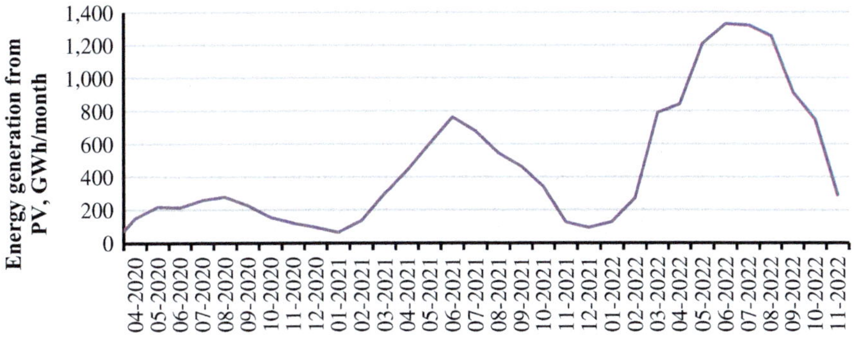

Fig. 2 Energy delivered from PV in Poland. Period 04/2020–11/2022 [9]

1.2 Literature Review

The economics of the energy storage process in current conditions was analyzed in [12, 13] even with the use of market energy prices, the profitability of energy storage is questionable. There are, of course, proposals for other forms of energy storage [14], including those within energy communities [15], but these solutions are not popularized in 2023.

The study [16] took into account the instability of a photovoltaic power plant depending on weather conditions and the possibility of annual electricity generation from photovoltaic power plants in the world and Ukraine.

In the case of prosumer PV micro-installations in Poland, households settle their bills on prosumer terms [17, 18] or net-billing [19–21], therefore the use of an energy storage in the current conditions cannot be economically profitable. Nevertheless, it positively affects the reduction of excess energy flows to the power grid [22] and helps in the pursuit of energy self-sufficiency of buildings [23, 24]. The use of an energy storage will also contribute to a greater possible share of Renewable Energy Sources in the national grid [25], which in turn may contribute to low-carbon development in Poland [4, 26].

There are studies on energy storage in mines, for example, inactive mine shafts have the potential to be used as energy storage [27]. Energy storage is an opportunity for the energy transformation in Poland [28] also as heat storage [29]. The search for new solutions in the field of energy saving [30] and the orientation of the energy economy towards a greener one is important for Poland [31]. Combustion of fossil fuels causes pollution, including contributes to the phenomenon of so-called "low emission" [32]. For this reason, in recent years, energy obtained from photovoltaics has been showing increasing interest in Eastern European counties [33].

2 Research Object

2.1 Photovoltaic Microinstallation in Poland

The object of the research is a prosumer micro-installation located in south-eastern Poland, it has been described in detail in the works on the use of rack for PV panels [34] and in the work on the verification of the quantitative production of energy from photovoltaics [35].

The installation consists of 14 PV panels (Fig. 3) with a total power of 5.04 kWp and an inverter with a power of 5 kW. The inverter measures the amount of energy supplied to the internal electrical installation. These data are collected in a 5-min interval.

Aggregated data on energy production up to an hourly interval is presented in Fig. 4.

The PV panels are directed towards the south-west, which means that the maximum daily values of the energy produced are observed in the afternoon (between 13 and 16).

Fig. 3 Photo of PV installation in Leki located in Lesser Poland. *Source* Piotr Olczak photo, used also in [36]

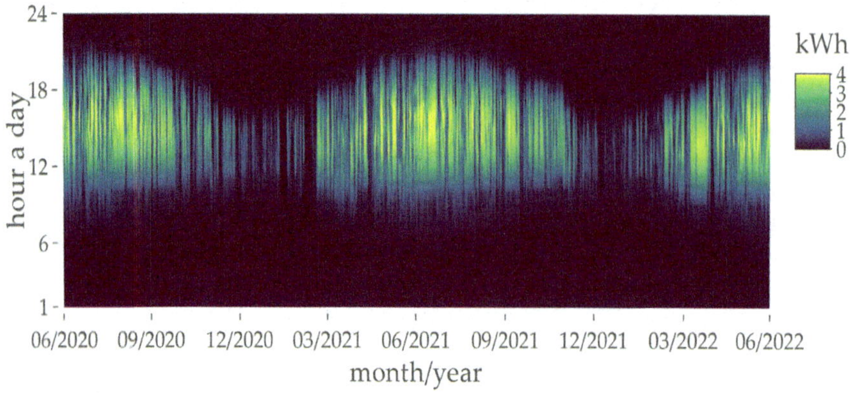

Fig. 4 Hourly values of energy produced in PV installation (EPV)

From the end of May 2020 (several days after the construction of the PV installation), a two-way energy meter was installed in the home electrical installation. It measures:

(1) Hourly values of energy received from the grid (ER).
(2) Hourly values of energy sent to the grid (ES).

Comparing the results presented in Figs. 4 and 5, one can notice the influence of the amount of energy produced on the reduction of the amount of energy taken from the grid. In addition, there is a noticeable increase in energy consumption in the autumn and winter periods.

Between 10/2020 and 1/2021, increased energy consumption at night was associated with the use of additional electric heating (Fig. 6.).

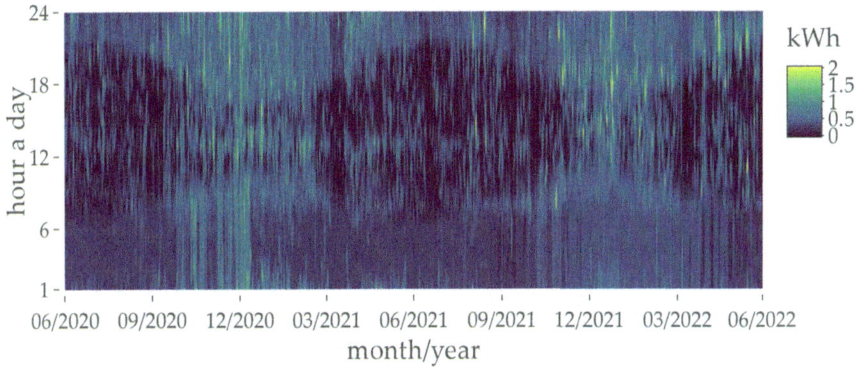

Fig. 5 Hourly values of energy received from the grid (ER)

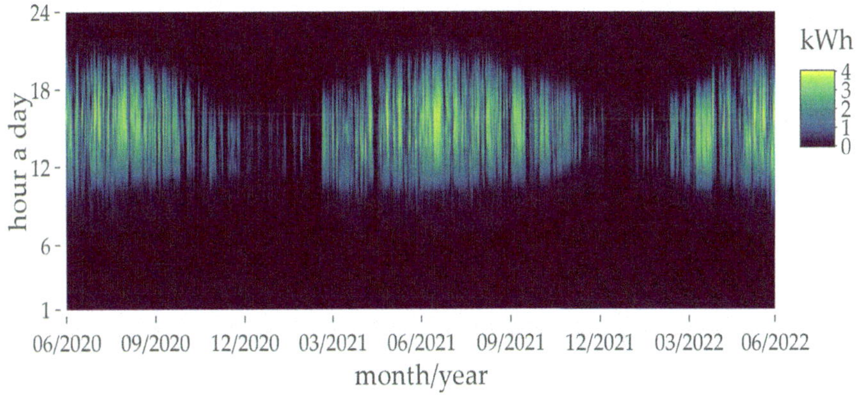

Fig. 6 Hourly values of energy sent to the grid (ES)

Table 1 Summary of results noticed in PV installation located in Leki (Lesser Poland), capacity 5.04 kWp

Parameter	1st year	2nd year	Unit
Start	1/06/2020	1/06/2021	
End	31/05/2021	31/05/2022	
Energy produced in PV installation (EPVy)	4671.8	5188.0	kWh/year
Specific yield	926.9	1029.4	kWh/kWp/year
Energy received from the grid (ER)	3398.3	3134.5	kWh/year
Energy sent to the grid (ES)	3235.8	3780.3	kWh/year
Energy self-consumed (ESCy)	1453.7	1407.7	kWh/year
Self-consumption rate	31.1	27.1	%
Energy consumed (EC)	4852.0	4542.2	kWh/year

The values shown in Fig. 4 reduced by the value of self-consumed energy (especially in the autumn and winter months and in the morning and evening) are the values of energy sent to the grid.

Table 1 gives a summary on an annual basis of recorded and calculated values (sum of values from Figs. 4, Fig. 5 and Fig. 6).

The observed year-on-year differences result from different weather conditions (mainly insolation [35]) and the variability of household energy use/consumption. This in turn is determined as: yearly sum of energy consumed (EC) was calculated by the following equation:

$$EC(year) = ESCy(year) + ER(year) \tag{1}$$

where: ESCy—yearly sum of energy self-consumed, kWh/year; ER—yearly sum of energy received from the grid, kWh/year.

2.2 *Energy Storage System*

Lithium-ion electricity storage was chosen as the energy storage technology. This solution is widely used in the Battery Energy Storage System [22, 36–38] and electromobility [36, 39]. The calculations take into account the usable capacity of the *ES0* energy storage, which is approximately 80% of the nominal/total capacity of the electricity storage. In the further part of the article, only *ES0* values were used.

3 Methodology

Calculations of the use of the electric energy storage were carried out—the following algorithm (Fig. 7).

Various usable capacities of the ES0 energy storage from 1 to 120 kWh are taken into account.

Calculated:

- ESS—current amount of energy stored in the energy storage in each hour τ, kWh,
- ESwS—the amount of energy sent to the grid when an energy storage is used, kWh,
- ERwS—the amount of energy received from the grid in the case of using an energy storage, kWh,
- ESSS—cumulative sum amount of energy stored in the energy storage, kWh.

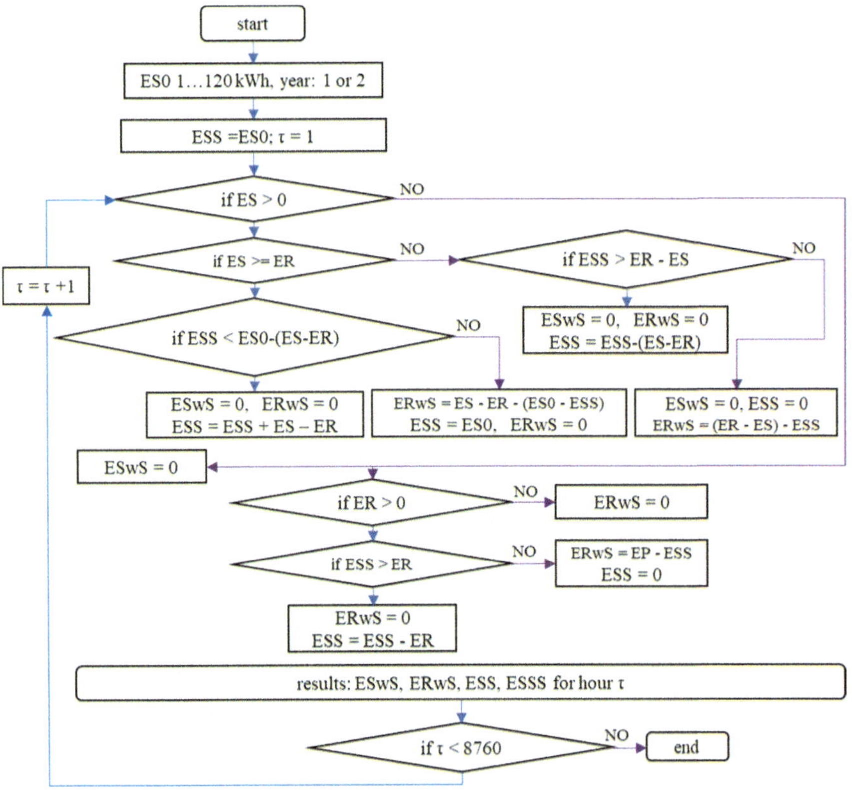

Fig. 7 Algorithm of calculation Energy sent to the grid with storage (ESwS) and Energy received from the grid with storage (ERwS). Based on installation results from 06/2020–05/2022; year 1: 06/2020–05/2021 and year 2: 06/2021–05/2022 [40]

Yearly sum of energy self-consumed was calculated by following equation:

$$ESCy(year, ES0) = EPVy(year) - ERwSy(year, ES0) \quad (2)$$

where: ESCy—yearly sum of energy self-consumed, kWh/year; EPVy—yearly sum of energy produced in PV installation, kWh/year; ERwSy—yearly sum of energy received in case of using energy storage, kWh/year; ES0—energy storage capacity, kWh.

Average yearly sum of energy self-consumed was calculated by following equation:

$$ESC2(ES0) = (ESCy(year = 1, ES0) + ESCy(year = 2, ES0))/2 \quad (3)$$

where: ESC2—average of yearly sums of energy self-consumed, kWh/year.

Yearly sum of energy stored was calculated by following equation:

$$ESSS2(ES0) = (ESSS(year = 1, ES0) + ESSS(year = 2, ES0))/2 \quad (4)$$

where: ESSS2—mean value of yearly energy stored, kWh/year; ESSS—yearly sum of energy stored, kWh/year.

Hourly value of energy self-consumed was calculated by following equation:

$$ESCh(\tau, year, ES0) = EPV(\tau, year) - ESwS(\tau, year, ES0) \quad (5)$$

where: ESCh—energy self-consumed in each hour, kWh; EPV—energy produced in PV installation, kWh; ESwS—energy sent to the grid in case of using energy storage, kWh.

4 Results

Based on method (Chap. 3) energy stored and energy self-consumed values were calculated—Fig. 8.

Based on Eq. 5 hourly value of energy self-consumed was calculated (Fig. 9).

The values of self-consumed energy occur only in the periods of energy production from photovoltaics (EPV—Fig. 4) in the case without energy storage. However, when we use an electricity storage, positive values of self-consumed energy occur practically in all hours of the year (except the autumn and winter season) (Fig. 10).

The case of using an energy storage with a capacity of 120 kWh is shown in Fig. 11. A slight difference in values can be seen between Fig. 10 and Fig. 11, which is also shown in Fig. 9: the difference of self-consumed energy 3157 (for ES0 = 9 kWh, self-consumed energy rate = 64%) a 3444 kWh/year (for ES0 120 kWh, value of energy produced in PV is higher than 4400 kWh/year and value of energy consumed is higher than 4500 kWh/year; self-consumed rate = 70%). This results

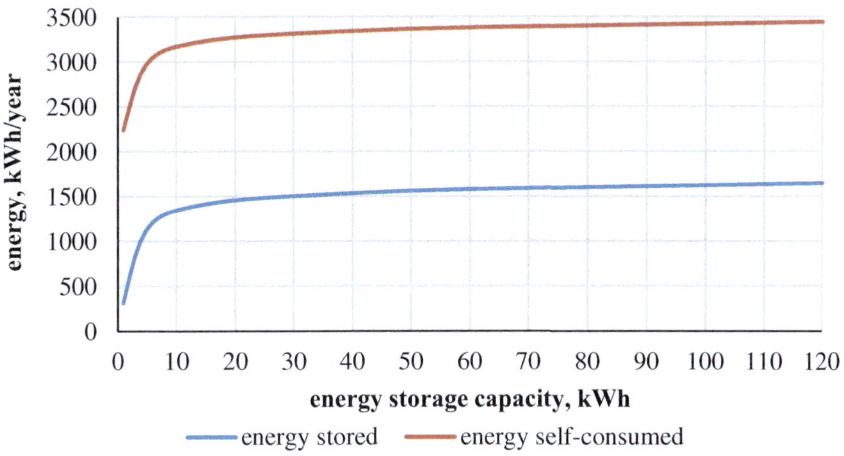

Fig. 8 Energy stored (*ESS2*) and energy self-consumed (*ESC2*) per year. Based on installation results from 06/2020–05/2022

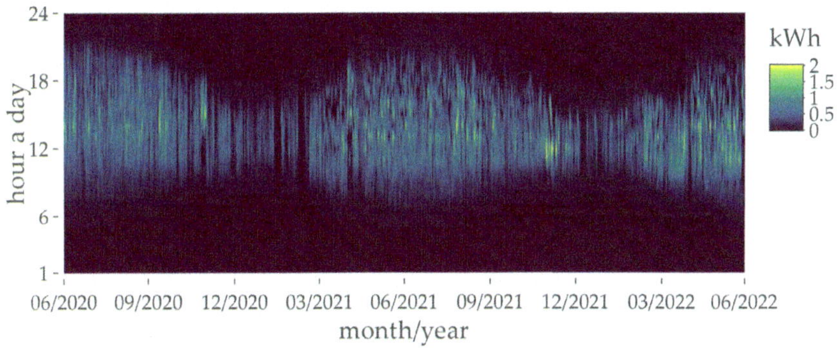

Fig. 9 Hourly values of self-consumed energy, without using energy storage

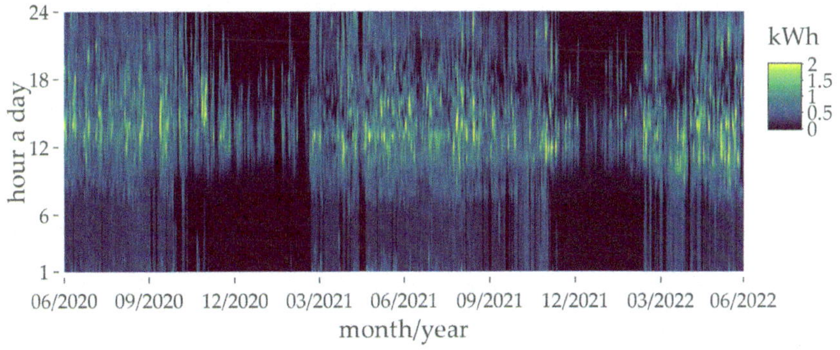

Fig. 10 Energy self-consumed per year in case of using energy storage capacity 9 kWh (ESCh(ES0 = 9 kWh))

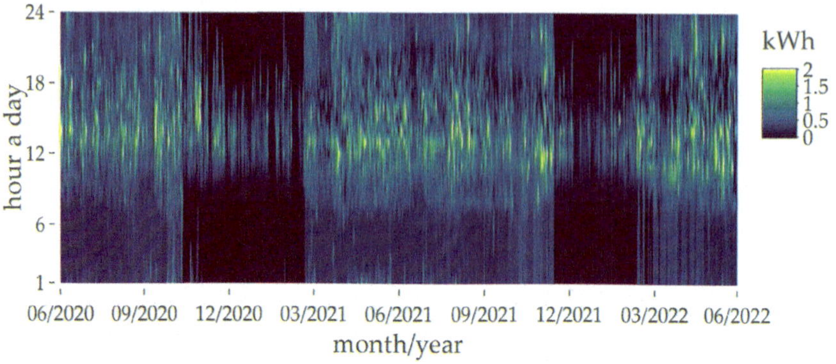

Fig. 11 Energy self-consumed per year in case of using energy storage capacity 120 kWh (ESCh(ES0 = 120 kWh))

in a lower degree of use of the storage infrastructure, expressed, for example, by the number of charging/discharging cycles per year, similarly to the work by Lepszy [40, 41].

In addition, it was hypothetically checked at what size of energy storage for the analyzed installation the self-consumption would be close to 100% per year. 1.7 MWh capacity reached. This means the cost of the energy storage of over EUR 1 million [12].

5 Conclusions

For a photovoltaic micro-installation located in Eastern Europe, calculations were made of the use of an electricity storage for the needs of balancing the energy production of a PV installation (household scale). For installations with a capacity of 5.04 kWp, the impact of using an energy storage with a capacity ranging from 1 to 120 kWh was analyzed based on two-year actual data recorded in the installation.

Energy storage with a capacity of 1–9 kWh can significantly increase the self-consumed energy from about 30 to 64% per year. Larger-capacity energy storage facilities only affect seasonal electricity storage (increasing the self-consumption rate by several percentage points). It was found that the use of an energy storage with a capacity of only 1.7 MWh can make a home photovoltaic micro-installation virtually energy self-sufficient. However, the investment costs of such a solution are over EUR 1 million, with annual household electricity costs of less than EUR 2000.

As part of further research, detailed economic calculations of the use of energy storage for prosumer PV micro-installations and the selection of a solution (size of electricity storage) are planned.

References

1. PTPiREE: Energetyka w liczbach mikroinstalacje. http://www.ptpiree.pl/energetyka-w-polsce/energetyka-w-liczbach/mikroinstalacje-w-polsce
2. Olczak P, Żelazna A, Matuszewska D, Olek M (2021) The "my electricity" program as one of the ways to reduce CO_2 emissions in Poland. Energies 14:7679. https://doi.org/10.3390/en14227679
3. Kulpa J, Olczak P, Surma T, Matuszewska D (2022) Comparison of support programs for the development of photovoltaics in Poland: my electricity program and the RES auction system. Energies 15:121. https://doi.org/10.3390/en15010121
4. Dzikuć M, Piwowar A, Dzikuć M (2022) The importance and potential of photovoltaics in the context of low-carbon development in Poland. Energy Storage Sav 1:162–165. https://doi.org/10.1016/j.enss.2022.07.001
5. Koval V, Sribna, Y, Gaska K (2019) Energy cooperation Ukraine-Poland to strengthen energy security. E3S Web Conf 132:1009. https://doi.org/10.1051/e3sconf/201913201009
6. Lee CY, Ahn J (2020) Stochastic modeling of the levelized cost of electricity for solar PV. Energies 13:3017. https://doi.org/10.3390/en13113017
7. Al-Khori K, Bicer Y, Koç M (2021) Comparative techno-economic assessment of integrated PV-SOFC and PV-battery hybrid system for natural gas processing plants. Energy 222:119923. https://doi.org/10.1016/J.ENERGY.2021.119923
8. Denysiuk S, Korotenko I, Lylo I (2019) Formation of network infrastructure of intellectual electricity communities in Ukraine. POWER Eng Econ Tech Ecol 2:7–16. https://doi.org/10.20535/1813-5420.2.2019.189982
9. PSE: Dane Systemowe. https://www.pse.pl/dane-systemowe
10. Rokicki T, Koszela G, Ochnio L, Perkowska A, Bórawski P, Bełdycka-Bórawska A, Gradziuk B, Gradziuk P, Agnieszka S, Szeberényi A, Dzikuć M (2022) Changes in the production of energy from renewable sources in the countries of Central and Eastern Europe. Front Energy Res 10. https://doi.org/10.3389/fenrg.2022.993547
11. Skłodowska M, Dopłaty w programie Mój Prąd 4.0 nie pomagają. Polacy nie kupują magazynów energii, https://wysokienapiecie.pl/71088-doplaty-w-programie-moj-prad-4-0-nie-pomagaja-polacy-nie-kupuja-magazynow-energii/
12. Komorowska A, Olczak P, Hanc E, Kamiński J (2022) An analysis of the competitiveness of hydrogen storage and Li-ion batteries based on price arbitrage in the day-ahead market. Int J Hydrogen Energy 47:28556–28572. https://doi.org/10.1016/j.ijhydene.2022.06.160
13. Nitsenko V, Mardani A, Streimikis J, Shkrabak I, Ivan K, Novomlynets O, Podolska O (2018) Criteria for evaluation of efficiency of energy transformation based on renewable energy sources. Montenegrin J Econ 14:237–247. https://doi.org/10.14254/1800-5845/2018.14-4.17
14. Sołtysik M, Kozakiewicz M, Jasiński J (2022) Improvement of operating efficiency of energy cooperatives with the use of "crypto-coin mining". Energies 15:8061. https://doi.org/10.3390/en15218061
15. Sołtysik M, Mucha-Kuś K, Kamiński J (2022) The new model of energy cluster management and functioning. Energies 15:6748. https://doi.org/10.3390/en15186748
16. Rubanenko O, Yanovych V (2020) Analysis of instability generation of photovoltaic power station. In: 2020 IEEE 7th international conference on energy smart systems (ESS), pp 128–133
17. Mikhno I, Redkva O, Udovychenko V, Tsimoshynska O, Koval V, Kopacz M (2022) Sustainable energy supply management in the mechanical-engineering industry. Polityka Energ—Energy Policy J 25:39–54. https://doi.org/10.33223/epj/156627
18. Wróbel J, Sołtysik M, Rogus R (2019) Selected elements of the neighborly exchange of energy—profitability evaluation of the functional model. Polityka Energ 22:53–64. https://doi.org/10.33223/epj/112508
19. Tongsopit S, Junlakarn S, Wibulpolprasert W, Chaianong A, Kokchang P, Hoang NV (2019) The economics of solar PV self-consumption in Thailand. Renew Energy 138:395–408. https://doi.org/10.1016/J.RENENE.2019.01.087

20. Trela M, Dubel A (2022) Net-metering versus net-billing from the investors perspective—impacts of changes in RES financing in poland on the profitability of a joint photovoltaic panels and heat pump system. Energies 15:227. https://doi.org/10.3390/en15010227
21. Roldán Fernández JM, Burgos Payán M, Riquelme Santos JM (2021) Profitability of household photovoltaic self-consumption in Spain. J Clean Prod 279:123439. https://doi.org/10.1016/J.JCLEPRO.2020.123439
22. Zelazna A, Gołębiowska J, Zdyb A, Pawłowski A (2020) A hybrid versus on-grid photovoltaic system: multicriteria analysis of environmental, economic, and technical aspects in life cycle perspective. Energies 13:3978. https://doi.org/10.3390/en13153978
23. Jurasz J, Canales FA, Kies A, Guezgouz M, Beluco A (2020) A review on the complementarity of renewable energy sources: concept, metrics, application and future research directions. Sol Energy 195:703–724. https://doi.org/10.1016/J.SOLENER.2019.11.087
24. Jurasz JK, Dąbek PB, Campana PE (2020) Can a city reach energy self-sufficiency by means of rooftop photovoltaics? Case study from Poland. J Clean Prod 245:118813. https://doi.org/10.1016/J.JCLEPRO.2019.118813
25. Sobczyk W, Sobczyk EJ (2021) Varying the energy mix in the EU-28 and in Poland as a step towards sustainable development. Energies 14. https://doi.org/10.3390/en14051502
26. Komorowska A (2021) Can decarbonisation and capacity market go together? The case study of Poland. Energies 14:5151. https://doi.org/10.3390/en14165151
27. Kulpa J, Kamiński P, Stecuła K, Prostański D, Matusiak P, Kowol D, Kopacz M, Olczak P (2021) Technical and economic aspects of electric energy storage in a mine Shaft—Budryk case study. Energies 14. https://doi.org/10.3390/en14217337
28. Dyczko A, Kamiński P, Stecuła K, Prostański D, Kopacz M, Kowol D (2021) Thermal and mechanical energy storage as a chance for energy transformation in Poland. Polityka Energ—Energy Policy J 24:43–60. https://doi.org/10.33223/epj/141867
29. Kuta M (2023) Mobilized thermal energy storage (M-TES) system design for cooperation with geothermal energy sources. Appl Energy 332:120567. https://doi.org/10.1016/j.apenergy.2022.120567
30. Stecuła K, Brodny J (2017) Generating knowledge about the downtime of the machines in the example of mining enterprise. In: International multidisciplinary scientific GeoConference surveying geology and mining ecology management, SGEM, pp 359–366
31. Kulpa J, Olczak P, Stecuła K, Sołtysik M (2022) The impact of RES development in Poland on the change of the energy generation profile and reduction of CO_2 emissions. Appl Sci 12. https://doi.org/10.3390/app122111064
32. Stecuła K, Tutak M (2018) Causes and effects of low-stack emission in selected regions of Poland. In: International multidisciplinary scientific GeoConference surveying geology and mining ecology management, SGEM, pp 357–364
33. Olczak P, Matuszewska D, Lishchenko A, Zhydyk I, Koval V, Iermakova O (2022) The economic efficiency of photovoltaic energy for energy prosumers. Polityka Energ—Energy Policy J 25:95–114. https://doi.org/10.33223/epj/155095
34. Koval V, Sribna Y, Kaczmarzewski S, Shapovalova A, Stupnytskyi V (2021) Regulatory policy of renewable energy sources in the European national economies. Polityka Energ—Energy Policy J 24:61–78. https://doi.org/10.33223/epj/141990
35. Olczak P (2022) Comparison of modeled and measured photovoltaic microinstallation energy productivity. Renew Energy Focus 43:246–254. https://doi.org/10.1016/j.ref.2022.10.003
36. Ostapenko O, Olczak P, Koval V, Hren L, Matuszewska D, Postupna O (2022) Application of geoinformation systems for assessment of effective integration of renewable energy technologies in the energy sector of Ukraine. Appl Sci 12. https://doi.org/10.3390/app12020592
37. Zhang J, Cho H, Luck R, Mago PJ (2018) Integrated photovoltaic and battery energy storage (PV-BES) systems: an analysis of existing financial incentive policies in the US. Appl Energy 212:895–908. https://doi.org/10.1016/j.apenergy.2017.12.091
38. Munzke N, Büchle F, Smith A, Hiller M (2021) Influence of efficiency, aging and charging strategy on the economic viability and dimensioning of photovoltaic home storage systems

39. Brodacki D, Polaszczyk J (2018) Emissivity of the operation of electric cars in the context of the strategic goals of the development of electromobility in Poland and the Netherlands. Polityka Energ 21:99–116
40. Olczak P (2022) Magazynowanie energii elektrycznej w prosumenckich mikroinstalacjach fotowoltaicznych. IGSMiE PAN, Kraków, Poland
41. Lepszy S (2020) Analysis of the storage capacity and charging and discharging power in energy storage systems based on historical data on the day-ahead energy market in Poland. Energy 213:118815. https://doi.org/10.1016/j.energy.2020.118815

Management Supply Chains Electric Vehicle Battery Recycling

Viktor Koval, Yevheniia Sribna, Olha Prokopenko, Liliya Filipishyna, Viacheslav Ivata, and I Wayan Edi Arsawan

Abstract The electric vehicle market—with an annual growth rate of more than 152% and reaching 75% of the market share of energy storage systems—poses qualitative and quantitative changes to the development of the market for electric vehicle battery recycling, where the recycling rate is only 7%. A related market for the production of electric vehicle batteries and the organization of logistics chains for the disposal of spent batteries is being formed, which functions with the state support of the electric vehicle industry. Recycling batteries takes place on an organizational and technological platform using pyrolysis, metallurgical and hydrometallurgical processes, and a combined variant of recycling electric vehicle batteries is also being developed. Such reorientation from simple processing is a consequence of environmental requirements and the economic need to offset the costs of rare and expensive metals partially. This study assesses the development of methods to use and recycle electric vehicle batteries under four scenarios, depending on state regulation and financing levels and on world exchange prices for non-ferrous and rare earth metals. The transition to new technological processing will allow repurposing up to 85% of the used metals. The problem of processing lithium batteries will become more acute in connection with the growing demand for lithium. Therefore, with a production

V. Koval (✉)
Izmail State University of Humanities, 12 Repina Str., Izmail 68610, Ukraine
e-mail: victor-koval@ukr.net

Y. Sribna
National University of Water and Environmental Engineering, 11 Soborna Str., Rivne 33000, Ukraine

O. Prokopenko
Estonian Entrepreneurship University of Applied Sciences, 10a Suur-Sojamae, 11415 Tallinn, Estonia

Sumy State Makarenko Pedagogical University, 87 Romenska Str., Sumy 40000, Ukraine

L. Filipishyna · V. Ivata
Admiral Makarov National University of Shipbuilding, 9 Heroiv Stalinhradu Ave, Mykolaiv 54025, Ukraine

I. W. E. Arsawan
Politeknik Negeri Bali, Bukit Jimbaran, Kuta Selatan 80364, Bali, Indonesia

V. Koval and P. Olczak (eds.), *Circular Economy for Renewable Energy*, Green Energy and Technology, https://doi.org/10.1007/978-3-031-30800-0_3

volume of 536.71 trillion electric vehicles, markets will reach equilibrium where lithium will be obtained through the complete recycling of batteries.

Keywords Electric vehicle industry · Supply chain sustainability · Recycling market · Energy consumption

1 Introduction

The era of automobile electric transport is a generally recognized trend in the greening of the transport sector and the development of a new market, and therefore serves as an indicator for private investors and provides them with a higher return on invested capital [1]. The process of alternative value is manifested when capital is directed to new markets [2].

As electric vehicles (EVs) grow in popularity, disposal of the vehicles and their parts becomes a more significant problem [3]. In developed countries, this problem is solved very simply—selling used EVs in developing countries [4].

From a global perspective, the need to recycle is an ever-growing problem for the EV industry and all industrial production. EV parts are highly specialized products made from rare and expensive materials and produced under strict environmental regulations; their disposal processes are specific. However, the specifics of the disposal of EVs have severe features [5]. They are the result of nanotechnological developments and include the mandatory use of rare and, accordingly, costly metals, such as lithium and cobalt [6]. Accordingly, the growth in demand for EV will likely lead to a massive increase in the volume of lithium. It could affect the use of lithium as a secondary material.

The rise of mass production and EVs sales, government stimulation of the industry and continuous improvement of the structure and technology of battery creation worsen the disposal problem. From a scientific point of view, this requires a deep and complete study and justification of such a problem [7, 8]. After all, the global scope of the problem requires the study of tools and their application in solving a whole set of interrelated disposal problems.

Accordingly, this study aims to assess the patterns and trends in utilizing the essential elements of EV construction following sustainable development and global environmental challenges.

Hypotheses

1. Environmental challenges force us to pay attention to the increased safety of using the latest rare elements in nanotechnology to create electric batteries for electric vehicles.
2. Economic factors (resource prices, profitability) cause the introduction of a full-fledged business focused on the disposal of elements of individual units and assemblies of electric vehicles.

2 Literature Review

Issues related to the use of automobiles are well documented [9–12]. Recent research has identified myths surrounding the purported environmental safety and efficiency of EV production and operation [13–15]. However, a detailed analysis of our research indicates that this conclusion is proper if the main component of EVs is not taken into account—the battery. Disposing batteries often leads to environmental disasters, manifested in self-ignition and problems of the long-term extinguishing of electric vehicles [16, 17].

Taking into account the dynamics of changing exchange prices of the primary metals used in electric vehicle production: cobalt, copper, lithium and nickel [18, 19].

The direction of scientific considerations that EVs will dramatically reduce environmental pollution and the level of exhaust gas emissions via the reduction in fuel and lubricant use [20–22] at first glance is rational and understandable. However, the process of disposing of high-tech batteries casts doubt on this proposition. Secondly, EV battery disposal uses harmful and dangerous chemical compounds. Thirdly, it emits dangerous substances such as salts and liquid compounds. Another aspect of this problem is that it is impossible to apply robotics or automation to these processes due to the lack of unified requirements for the structure and number of elements used to assemble storage batteries.

Numerous studies of battery recycling technologies describe their problems and note the impossibility of complete recycling [23–26]. At the same time, the emphasis is on technological solutions Numerous studies of battery recycling technologies describe their problems and note the impossibility of complete recycling [23–26]. At the same time, the emphasis is on technological solutions [27–29]. However, the problem of mass industrial recycling must be addressed as a single organizational and technological complex. This aspect will become more difficult with the increase in car production and the retirement of batteries. Therefore, the way out of such a situation is the formation of full-fledged logistics chains that carry out acceptance, temporary storage and accumulation of EV batteries with the application of bar marking to allow for automation and robotization of the direct process of material processing. This marking and automation technology is significant but has yet to be fully developed, and only China currently uses this approach [30].

Existing research forecasts the environmental friendliness and impact of EVs but does not consider recycling or the technologies used in recycling [31]. However, considering the state administration's priority for the complete electrification of modern transport, forecasting the future of EV battery disposal and recycling should be prioritized. Such processes are quite complex, considering the factors and the need for complete information on replacing used batteries and battery accumulation. Therefore, it is necessary to use the scenario method of forecasting, which helps determine the main directions of conceptual recycling and processing electric vehicle batteries.

3 Research Methodology

The evaluation and development of the EV market were carried out using dynamic indicators that characterized the change and dynamics of market development levels between 2016 and 2021. This period was selected because in 2017, the total number of EVs in operation, including hybrid electric vehicles, exceeded 3 million.

Economic analysis of changes in EV sales volume was carried out using the following indicators: absolute growth, coefficient (rate) of growth, and growth rate. These methods allowed the assessment of changes in the levels of indicators in relative terms, using both the chain method and the base method.

Plan-fact analysis was used to compare the actual global EVs sales volume to the forecasted volume. It made it possible to determine consumers' real interest in EVs and assess the scale of production and the dynamics of the business development.

In addition, in assessing environmental problems associated with the production, operation and disposal of electric vehicles, the following methods were applied:

1. Systematization, generalization of scientific publications on the assessment of the use and reuse of car batteries, as well as existing and developing methods of their processing;
2. Systematic and logical analysis, a method of information synthesis, which made it possible to carry out analytical comparisons and reconciliation of various trends and to present logical structures in the course of solving problems of disposal of used EVs and their units;
3. A logical analysis of the development of entrepreneurship in the field of recycling EV batteries made it possible to form a system of factors that determine the efficiency and economic feasibility of the secondary use of materials from EVs.

Studies of internal cause-and-effect, structural–functional direct and reverse relationships were conducted to produce logistics chains for the entire recycling cycle of EV batteries. In addition, it was used to highlight the main trends in scientific developments in the formation of the replacement of lithium and cobalt with other, more resistant elements to ignition and the temperature regime of battery operation. Due to the applied induction, generalizations and theoretical conclusions were formed. Regarding the applied dialectical method, the disposal of electric vehicle batteries was considered simple to modern technological processes. That is, there has been a change from simply throwing used EV batteries in the trash to forming an entrepreneurial case and business around their disposal. These processes are continually being improved and expanded. The study focuses on the qualitative parameters of the studied categories "cathode", "anode", and "lithium-ion batteries." The applied comparison method made it possible to evaluate the processes of disposal of EV batteries by world region.

Applied differential and difference equations to study the disposal process of electric vehicle batteries. I define the following mathematical model:

$$\begin{cases} \mathrm{Eecon_0} = \mathrm{Eecon_i(t)}; \\ \mathrm{Eecol_0} = \mathrm{Eecol_i}(t); \\ \mathrm{Etech_0} = \mathrm{Etech_i}(t); \end{cases} \tag{1}$$

The solution of the system of equations gives the following phase state:

$$\mathrm{k_n} = \mathrm{k(t_n)} = ((\mathrm{Eecon_0})_\mathrm{n}, (\mathrm{Eecol_0})_{\mathrm{n},} (\mathrm{Etech_0})_\mathrm{n}) \tag{2}$$

The limit solution of the system of equations is reduced to the system:

$$\begin{cases} \lim\limits_{t\to\infty} \mathrm{Eecon_0(t, (Eecon_0)_0)} = \mathrm{Eecon}_0^*(t) \\ \lim\limits_{t\to\infty} \mathrm{Eecol_0(t, (Eecol_0)_0)} = \mathrm{Eecol}_0^*(t) \\ \lim\limits_{t\to\infty} \mathrm{Etech_0(t, (Etech_0)_0)} = \mathrm{Etech}_0^*(t) \end{cases} \tag{3}$$

where $\mathrm{Eecon}_0^*(t)$, $\mathrm{Eecol}_0^*(t)$, $\mathrm{Etech}_0^*(t)$-limit value functions.

The simulations indicate that the disposal process can reach an ideal state over a period of time. However, in practical application, such an achievement is conditional. After all, numerous factors must be considered, especially those under commercial secrecy regarding the quantitative structure of the rare earth metals used in the cathode and electrolyte.

The study uses a predictive method called the scenario-building method to assess future trends in processing used EV batteries due to the determined sequence of changes and expansion of technologies for EV battery disposal and changes in global trends with increasing ecological dominance. The methodological basis of the research is the information and analytical materials of the relevant research and laboratory institutions, which are focused on the development and implementation in the production of charging devices, not only for EV production but also for devices for other modern electronics. All available methods used in business and state regulation research were used.

4 Results

The EV market is dynamically developing, which is confirmed by yearly sales volumes exceeding their forecasted values by 3.4% on average between 2016 and 2021. The exception is in 2021 when sales volumes 1.35% lower than forecast due to the change in production conditions brought on by the Covid-19 pandemic. At the same time, the worst situation occurred in 2021, when actual sales were 4207.65% below projected sales (Table 1).

In the battery EVs segment, the projected sales volumes have become underestimated. It suggests that consumers prefer pure EVs to hybrid EVs. The following factors contribute to the development of EVs at the global level:

Table 1 Assessment of actual and forecast changes in the volume of sales of electric vehicles, 2016–2022

Indicators	2016	2017	2018	2019	2020	2021
Total forecast	761.6	1206.9	1992.5	2186.1	3134.2	6507.9
Volumes of sales of electric vehicles, thousands of vehicles	792	1263	2062	2264	3240	6421
Difference	30.4	56.1	69.5	77.9	105.8	– 86.9
Deviation, %	3.84	4.44	3.37	3.44	3.27	– 1.35
Battery Electric Vehicles						
Volumes of sales of electric vehicles, thousands of vehicles	741.312	1178.379	1919.722	2096.464	3013.2	5977.951
Forecast	477.7	801.6	1367.8	1624.6	2165.4	4599.4
Difference	263.612	376.779	551.922	471.864	847.8	1378.551
Deviation, %	35.56	31.97	28.75	22.51	28.14	23.06
Plug-in Hybrid Electric Vehicles						
Volumes of sales of electric vehicles, thousands of vehicles	50.688	84.621	142.278	167.536	226.8	443.049
Forecast	283.8	405.3	624.7	561.5	968.8	19,085
Difference	– 233.112	– 320.679	– 482.422	– 393.964	– 742	– 18,642
Deviation, %	– 459.90	– 378.96	– 339.07	– 235.15	– 327.16	– 4207.65

Source Calculation based on [32–35]

1. Environmental factors:
   - fight against carbonization, i.e., pollution by emissions of harmful substances, exhaust gases from cars and heavy elements after combustion;
   - the growing popularity of sustainable development, which drives the introduction of environmentally friendly products.
2. Economic factors:
   - a growing market, therefore attractive to investors, as the profits from investments, will be due to increased sales volumes;
   - the constant reduction in the cost of EVs, which expands the pool of potential consumers.
3. Infrastructure factors:
   - implementation of relevant developments regarding fast and reliable EV charging through appropriate charging stations;
   - setting up the service and maintenance of EVs through appropriate service stations.

4. Technological factors:
   - constant improvement of structures and units of cars, which ensure an increase in mileage on one car charger;
   - technological implementations to increase the capacity of storage batteries.
5. State regulation:
   - replacement of 50% of motor vehicles with electric vehicles by 2030 in the world;
   - state subsidies for the purchase of EVs from developed countries.

From 2016 to 2021, global sales volumes grew dramatically. The growth rate reached more than 100% by 2018. Since 2019, the growth rate has approximately doubled yearly. For battery EVs, the growth rate using 2016 as the base year closely corresponds to the general growth of all-electric vehicles, but with a slight lag. Hybrid electric vehicles have a high growth rate close to battery EVs. Again using 2016 as the base year, the sales growth rate was higher for hybrid EVs than battery EVs. Their low absolute sales volume explains this in 2016 (about 50,000 vehicles) (Table 2).

In general, the lowest growth rate occurred in 2019. However, the average annual growth rate for each segment and the total reached 1.05, and the growth rate exceeded 152.0%. The ecological approach in estimating the emissions of electric vehicles is equated to that of conventional cars. Therefore, a paradoxical system emerges: during the operation of an EV, carbon dioxide emissions amount to 27.1 tons, as opposed to 22.8 tons for ordinary cars. The harmful effects of the batteries themselves are not taken into account, but it is the batteries that carry the leading environmental threat. After all, they are produced based on nanotechnology, and the corresponding processing system has yet to be formed. Technologically, only the usual processing of lead-acid batteries has been developed (Table 3).

Significant growth in demand for EVs requires a corresponding increase in the use of various metals. Modern EVs mainly use copper, nickel, graphite, metals of the platinum group, lithium, and cobalt, as well as rare earth elements: lithium, cobalt and graphite are critical components of modern batteries. So, depending on the model, producing one modern EV requires 12 to 50 kgs of lithium, 5 to 8 kgs of nickel, and about 54 kgs of graphite.

Nickel is used in the production of batteries and makes up 80% of them [36]. It causes an increase in demand for the metals mentioned above. Thus, lithium, cobalt, nickel, and copper prices increased by 60%, 250%, 26%, and 49%, respectively, between 2016 and 2022 [37]. A helpful comparison is that during the same period, the value of gold on the stock exchange increased by 16% and silver by 17% [37]. Lithium is a critically important element in the production of electric storage batteries. Experts forecast that a 1% increase in EVs production increases the annual demand for lithium by 70,000 tons. According to forecasts, the number of cars will reach 2 billion by 2050. Half of them will be electric cars, which use 9 million tons of lithium and 13 million tons of cobalt [33]. We can evaluate the use of lithium in the production of electric batteries. We do this by determining the mathematical dependence of the

Table 2 Statistical calculation of the dynamics of sales of electric vehicles by segments, 2016–2021

Indicators	2016	2017	2018	2019	2020	2021
Total						
Volumes of sales of electric vehicles, thousands of vehicles	792	1263	2062	2264	3240	6421
Annual absolute growth	–	471	799	202	976	3181
Growth factor compared to the previous year	–	1.59	1.63	1.10	1.43	1.98
Growth rate	–	159.47	163.26	109.80	143.11	198.18
Growth rate (chain)	–	59.47	63.26	9.80	43.11	98.18
Growth rate (baseline)		159.47	260.35	285.86	409.09	810.73
Battery electric vehicles						
Volumes of sales of electric vehicles, thousands of vehicles	741.31	1178.38	1919.7	2096.46	3013	5977.95
Annual absolute growth	–	437.067	741.34	176.742	916.7	2964.75
Growth factor compared to the previous year	–	1.59	1.63	1.09	1.44	1.98
Growth rate	–	158.96	162.91	109.21	143.73	198.39
Growth rate (chain)	–	58.96	62.91	9.21	43.73	98.39
Growth rate (baseline)		158.96	258.96	282.80	406.47	806.40
Plug-in hybrid electric vehicles						
Volumes of sales of electric vehicles, thousands of vehicles	50.688	84.621	142.28	167.536	226.8	443.049
Annual absolute growth	–	50.688	84.621	142.278	167.5	226.8
Growth factor compared to the previous year	–	1.67	1.68	1.18	1.35	1.95
Growth rate	–	166.94	168.14	117.75	135.37	195.35
Growth rate (chain)	–	66.94	68.14	17.75	35.37	95.35
Growth rate (baseline)		166.94	280.69	330.52	447.44	874.07

Source Calculation based on [32–35]

production and sale of electric cars with the mathematical dependence of lithium costs in monetary form for producing electric batteries (Table 4).

The mathematical dependence of these quantities will be determined through the trend line (Fig. 1).

Mathematically, we equate the trends and find out the volume of sales of cars in which lithium will be used from the processing of electric batteries. Solving the quadratic equations for the supply of electric cars:

$$y = -17.645x2 - 1124x + 41686; \; y = 882.59x2 - 4188.6x + 4966.9$$

And lithium supply:

Table 3 Characteristics of the most common industrial electric batteries

Type of batteries	Producer	Composition of metals	The composition of the electrolyte
Lithium-nickel-manganese-cobalt-oxide battery	Gigafactory-1	Lithium, iron, manganese, cobalt and their compounds with phosphorus, oxygen	Ethylene carbonate, propylene carbonate, dimethyl carbonate
Lithium-nickel–cobalt-aluminum-oxide battery	Gigafactory-1	Lithium, iron, aluminum, cobalt and their compounds with phosphorus, oxygen	Ethylene carbonate, propylene carbonate, dimethyl carbonate, fluoroethylene carbonate

Table 4 General characteristics of lithium expenditures concerning sales of electric vehicles, 2016–2021

Years	Number of sold electric vehicles, million units	Lithium, USD per kg	The average price of an electric vehicle, USD	Use of the amount of lithium per sold electric vehicle, kg	Expenditures on lithium for the production of electric batteries, USD
2016	0.792	26.17	41,000	22.18	580.35
2017	1.263	27.39	39,000	35.36	968.62
2018	2.062	41.13	38,330	57.74	2374.68
2019	2.264	30.89	37,667	63.39	1958.18
2020	3.24	30.41	35,500	90.72	2758.80
2021	6.421	75.18	34,500	179.79	13,516.46

Source Compiled by the authors on [35, 38]

$$y = 88259\,x2 + 41886.6x + 49,\ 66.9 = -17.643x2 - 1124x + 41868$$

Setting these equations equal gives a production volume of 536.71 trillion electric cars. This volume will become a condition when lithium is used to process electric batteries. As of 2021, there is a trend toward increased demand for cobalt for the production of EVs, which has exceeded the demand for metal for the production of smartphones and computers. At the same time, the total demand for cobalt was 175 thousand tons, and its extraction reached 160 thousand tons. Therefore, according to these forecasts, by 2026, half of the global demand for cobalt will be from EVs manufacturers [39].

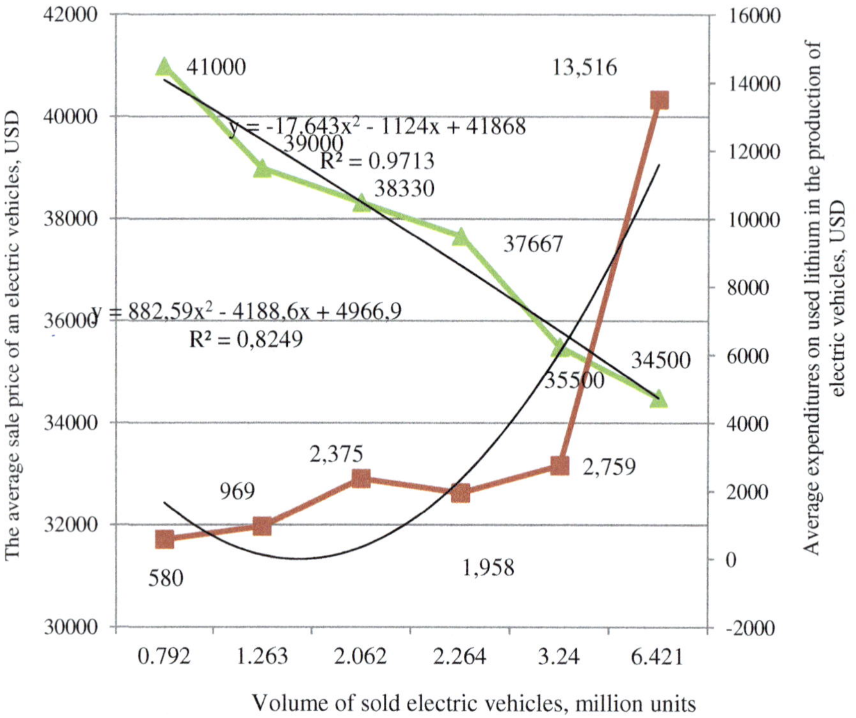

Fig. 1 Average expenditures on used lithium in the production of electric vehicles, USD

Cobalt is the most problematic material used in batteries. In operation, there are frequent cases when both mobile phones and electric vehicles catch fire.

It is not related to operating conditions but to technical deficiencies in producing this metal. In addition, lithium and cobalt enter the groundwater, posing a danger to the environment.

Lithium-ion batteries' efficacy depends on their external environment; namely, at sub-zero temperatures, they lose their effectiveness. A dangerous problem with Tesla cars is the rapid ignition of batteries, which occurs both when parked and in motion. There were fixed fatalities when a car caught fire after hitting a tree. It was possible to extinguish the fire only after four hours. Thus, another problem is the safety of firefighters, who must work in chemically aggressive environments when extracting the results of burning lithium and cobalt. In the summer, the risk of catching fire increases tenfold [40].

By 2010, most used batteries were sent to the Congo and Nigeria [41], which were mass incinerated in landfills. Currently, there is a phenomenon where landfill sites are filled with relatively new electric vehicles, for example, the EV-1 landfill in America [42]. In China, such areas are used for car-sharing vehicles, such as the Lifan 330 eV [43]. Such disposal is a consequence of the bankruptcy of the respective manufacturers and lessees of electric vehicles.

So, the main environmental problem of electric vehicles is the recycling of batteries. For lead-acid batteries, technology has been developed to isolate the components of lead, sulphuric acid, and lead oxide, producing so-called lead paste and lead acid. They are easily neutralized, and the materials are involved in re-production. For lithium-cobalt batteries, such technologies are relatively primitive and do not ensure the complete recycling of all elements.

According to this study's calculations, with the production of 536.71 trillion EVs, there will be a need for the complete disposal of electric batteries. However, for the current period, the main problem in recycling is the proportional composition of these elements, which takes more work to take into account in the recycling process. In addition, the labeling of lithium batteries itself needs to indicate complete information about the possibility of their disposal (the specifics of the disposal of the cathode, anode and electrolyte are not indicated). It is why the widespread current technology for the disposal of such batteries is grinding and pyrometallurgy due to its economic feasibility.

The development of chemical production and, accordingly, the automobile industry, which directly depends on the volumes of the chemical industry for the supply of plastics, rubber and others, have seriously damaged the planet. Various artificial polymers decompose over centuries, slowly poisoning the biosphere, accumulate in urban landfills, or are buried, leading to soil and groundwater pollution. In the modern era, all of the past solid household and industrial waste processing methods fail to guarantee their safe processing. Numerous harmful gases, particularly carbon dioxide and nitrogen, are released, increasing ecological danger [44]. However, an innovative approach to recycling is being implemented, focusing on using the pyrolysis method in waste processing. In other words, all waste except bricks, stones, and metal is processed at high temperatures of 1600–1800 °C (273 K) and high pressure. This type of processing produces slag or neutral residues. The obtained synthetic gas is a raw material that can produce methanol, higher alcohols, nitrogen fertilizers, ammonia, synthetic motor oil, and fuel.

Another approach involves high-temperature processing, resulting in syngas containing 90% hydrogen and carbon and 10% solid slag. Slag is created in the form of sand, silica, alumina and quicklime. A pyrometallurgical process has been formed, the concept of which assumes that after crushing the batteries, they are subjected to high temperatures. Then a carbon mass of metal, plastic and various viscous substances forms, which later, due to chemical and physical processes, allow their oxides and salts. In general, the pyrometallurgical process is an ore-processing concept. This approach counts batteries as received ore without incorporating extraction costs. This process allows processing any battery without considering related production technologies and the specifics of the structural composition, but it has high energy consumption.

A hydrometallurgical process has also been worked out. Its essence boils down to the fact that various acids act on the battery, forming a so-called liquid substance saturated with soluble metal salts that are part of the battery structure. This process is more energy efficient but dangerous. Its advantage is that it can obtain graphite and lithium, which are lost during pyrolysis. Using chemical compounds and acids causes is dangerous for workers and production.

These two approaches are somewhat independent. The practice of application combines both approaches while prioritizing one. Experimental developments regarding introducing new concepts are based on the formation of a direct recycling process initiated by the US National Renewable Energy Laboratory. The technological process's essence lies in removing electrolytes from batteries using a vacuum. The bound components are then extracted through crushing and heating and the action of solvents. A flotation technique is then used to separate the anode and cathode materials. At the output of processing, a powder-like material is obtained. This technology has been developed in laboratory conditions, but in practice, some difficulties arise:

1. The lack of uniform standard markings that would indicate the composition of the battery based on the chemical and quantitative characteristics of the metals used;
2. The lack of a standard method of disassembling the battery (current methods of battery disassembly require at least two hours);
3. The lack of unification of batteries prevents the use of robots to speed up disassembly costs;
4. The use of chemical solvents, which are prohibited for use in many countries, primarily European countries and the U.S.;
5. Powerful manufacturers of electric vehicles invest the most funds in the development of batteries, and they want to keep their knowledge private. If they transfer licenses, it means that the technologies are already outdated.

Therefore, the solution to this problem is carried out in three stages:

1. Replacement light metals (lithium, cobalt and others) with other materials. That is, new lithium-silicon and solid-state battery technology are being developed. Silicon is added to the battery composition (less than 10%), increasing service life and fire resistance while ensuring satisfactory charging speed. It reduces the number of cells in the battery, reducing the risk of failure. In addition, such batteries retain 90% of their charge capacity after 5000 cycles and charge up to 80% of battery capacity in 15 min.
2. The formation of technologies based on lithium-iron-phosphate batteries. By the end of 2021, there was a working prototype of such batteries. The peculiarity of this battery is that it can withstand 60-degree heat and cools down quickly. However, such batteries are still focused on sports EVs in mass application.
3. The development of sodium-ion battery technologies shows resistance to self-ignition and has a longer service life compared to lithium-ion ones. Their disadvantages are low charging speed and low charge density.

Currently, intensive laboratory studies are being conducted by EV manufacturers to change the structure of metals in electric batteries to increase their efficiency in use and reduce energy consumption during production and disposal. The main areas of development to increase the efficiency of batteries are presented in Table 5.

Table 5 Prospects for the development of electric batteries

Company	Battery characteristics	Mileage of an electric vehicle per charge, km	Production
SVOLT	Without cobalt content	Up to 600	Mass production has not started
Tsinghua	Solid state batteries	500, battery life cycle 10 million km 50 years of operation	Toyota and Nissan released an electric vehicle with such a battery
Samsung	Graphene batteries	5	Patent application

Source Compiled by the authors [45, 46]

Currently, a full-fledged battery recycling business is being formed. It has the national character of the producing country. The primary producers are diversifying rare earth metals. Therefore, battery recycling processes are oriented within the boundaries of a particular country. At the same time, such a diversification system allows manufacturers to attract state funding. It is explained by the possibility of attracting state support in the context of sustainable development, particularly the strengthening of greening (Table 6).

American companies such as Ionic Materials, QuantumScape, Sila Nanotechnologies, Sion Power, and Sionic Energy are already developing next-generation battery technologies [52].

The widespread introduction of factory battery recycling indicates that car manufacturers want to keep EV sales revenue within the company. Thus, they plan to avoid paying environmental taxes in the future. In a macroeconomic approach, the battery recycling business can only work if the value of lithium is exceptionally high. According to the London Metal Exchange [53] (LME) and Fastmarkets [54], in 2021, the price of lithium fell to 80.25 USD per kilogram (80,250 USD per ton). According to Morgan Stanley forecasts, lithium prices will fall by 30% by 2025 [55].

However, China is targeting mass recycling of car batteries by 2030 (recycling 120 tons will ensure the assembly of 200,000 new cars). Therefore, future battery recycling will be financed with indirect environmental fees. However, with state support, such an endeavor by the private sector is much better.

The modern direction of electric battery recycling is demonstrated by Chinese companies, in which a closed cycle of the recycling process is followed, from the collection, sorting, and storage to complete recycling with the reuse of the received battery elements. Thus, the construction of a full-fledged logistics chain of a closed cycle of complete processing of electric vehicle batteries, oriented from the consumer to the manufacturer of electric vehicles, is being formed. In addition, CATL operates in Southeast Asia, which provides 32.5% of the global output of batteries for electric vehicles, and two more manufacturers (LG Energy Solution and Panasonic), which together produce 70% of the battery market for electric vehicles (21, 5% and 14.7%, respectively) [52].

The logistics chain of a closed cycle of complete recycling of mobile electric batteries is the most advanced organizational complex for the disposal of used

Table 6 The modern process of processing mobile electric batteries

Company	Region	The essence of battery recycling	The reality of the project	Potential for recycling batteries
Volkswagen	Germany	Entire processing with the possibility of obtaining metals	Effective from 2020	The capacity of 3000 batteries or 1500 tons per year
Veolia	England	Complete rework	Implemented in 2021	By 2023, the amount of recycling of 20% of used batteries in the entire English market
			By 2024, another plant in Manward	Half of the market for used batteries
Renault	France	Reusing 90% of battery content methods	Research line in 2025	Will reduce greenhouse gas emissions during battery production by 50%
Duesenfeld	Germany	Metal salts (cobalt sulfate, lithium carbonate, nickel sulfate, manganese sulfate)	Valid since 2019	70% recycling of all used batteries in Germany
ReCell	China	Extraction of ready-made cathode crystals	Methodology development	Transition to the new principle of battery recycling by 2027
Umicore	EU, South, North America, Asia, Africa	70% recovery of cobalt, copper and nickel (loss of lithium)	Valid since 2020	The world leader in the processing of secondary precious and industrial metals It owns 46 processing enterprises around the world

Source Compiled by the authors [47–51]

batteries. The complex consists of two processing enterprises created depending on regional needs and located in administrative areas where EVs are sold. The first type is small recycling centers of the receiving type, where batteries are accumulated and stored with capacities of up to several tons of batteries. The second type is concentrated storage enterprises, where batteries are stored for a long time. They are located in centers where more than 8000 electric vehicles are concentrated and serviced. It is how the processing logistics chain is formed when the batteries go directly to the processing enterprises of large companies producing EVs from the receiving and storage points.

Another approach in forming the logistics of the battery plug is a narrowly specialized logistics chain. It involves unidirectional specialization of companies sorting used electric batteries according to the level of wear and tear. Such a business is established by the 4R Energy enterprise and the companies Nissan and Sumitomo. These links of the logistics chain work for narrow segments of the respective parent companies of EV manufacturers.

The main environmental problem caused by EVs is the recycling of electric batteries. According to forecasts of the National Renewable Energy Laboratory [54], the number of batteries produced will increase dynamically as a result of the expansion of the production of EVs. Therefore, the challenge of recycling them will grow over time. There are two types of solutions. One solution is private and public funding of research into the possibilities of recycling batteries based on innovative approaches. It is typical for large companies producing EVs, which evaluate capabilities based on their research laboratories. The second solution is forming a full-fledged battery recycling market as a derivative of the EV market.

All the analyzed problems indicate that the economic feasibility of switching to electric transport, which the governments of leading countries support, needs to be measured more. It can be achieved only through administrative coercion. Since such a mechanism is not used in market conditions, a tax on the disposal of EVs will be introduced sooner or later. However, even with additional funding, it will not be possible to achieve effective processing due to restrictions on labor protection. Therefore, there will be a trend toward accumulating battery landfills in developing countries. At the same time, given the constantly growing demand for rare earth metals, EVs manufacturers will try to cover them at the expense of their processing.

Factors influencing the industrial disposal of EVs batteries noted in the study can be grouped to estimate a mathematical dependence on these processes.

1. The dependence determines the economic factor: the increase in prices for rare earth metals, the greater the economic expediency of using materials reuse. For the current period, the economic factor makes up 45% of the metal used (including chemical salts and solutions).

   Economic factor : $\mathrm{Econ}_0 = \mathrm{Econ}_i(t)$ (4)

2. The environmental factor is defined as universal and responsible. All developed countries have taken a course to switch transport to an electric type of engine.

$$\text{Environmental factor} : \mathrm{Ecol}_0 = \mathrm{Ecol}_i(t) \quad (5)$$

3. The technological factor is the fundamental physical, chemical and nano-developments used in the modern recycling electric vehicle batteries.

$$\text{Technological factor} : \mathrm{Etech}_0 = \mathrm{Etech}_i(t) \quad (6)$$

In the mathematical dimension, this is the solution of a system of equations, which the corresponding indicators give at each moment of time t. At a certain moment in time, this system will be determined by the vector of phase coordinates (Formulas 4–6).

Then the trends in the change of the recycling process are differential equations that reflect the change in the rate of change of a separate phase (namely, the phase of the development of recycling processes) through the function of the phase coordinate and time for each of the three equations (Formulas 1–3).

As a result, the following differential equations are obtained:

$$\begin{cases} \frac{d(\mathrm{Eecon}_0)}{dt} = \mathrm{Eecon}_0 = f(\mathrm{Eecon}_i(t), \mathrm{Eecol}_i(t), \mathrm{Etech}_i(t), t) \\ \frac{d(\ \mathrm{Eecol}_0)}{dt} = \mathrm{Eecon}_0 = f(\mathrm{Eecon}_i(t), \mathrm{Eecol}_i(t), \mathrm{Etech}_i(t), t) \\ \frac{d(\mathrm{Etech}_0)}{dt} = \mathrm{Etech}_0 = f(\mathrm{Eecon}_i(t), \mathrm{Eecol}_i(t), \mathrm{Etech}_i(t) \end{cases} \tag{7}$$

For this model, the initial state k_0 at (t = 0, the start of the time count) is determined by the values of the phase coordinates at time t0. Therefore, the state of the system at this initial time will be the equation:

$$k_0 = k(t_0) = ((\mathrm{Eecon}_0)_{t_0},(\ \mathrm{Eecol}_0)_{t_0},(\mathrm{Etech}_0)_{t_0}) = ((\mathrm{Eecon}_0)_0,(\mathrm{Eecol}_0)_{0,}(\mathrm{Etech}_0)_0) \tag{8}$$

and the final state will be determined by the reference values of these system components.

Thus, the proposed model of EV battery disposal is a system of differential equations where the three essential components of the concept, which are also factors in the disposal process, are unknown. However, over time, this system will be transformed into a system that a limiting value can describe (at $t \to \infty$, and initial values $((\mathrm{Eecon}_0)_0$, $(\ \mathrm{Eecol}_0)_0$, $(\mathrm{Etech}_0)_0$.

Therefore, the system we built has an attractor in the form of a limit cycle. In summary, there will always be such a value $t < \infty$ at which the limit value functions $\mathrm{Eecon}_0^*(t)$, $\mathrm{Eecol}_0^*(t)$, $\mathrm{Etech}_0^*(t)$ will reach the indicators of the given ideal value of the components of electric vehicle battery disposal determined by us. Therefore, the disposal process is implemented in a particular finite time. The time will be achieved under centralized control over the maximization of the levels of obtaining the used metals. In other words, the calculation notes that processing is possible in the minimum time, but this time is determined by the given level of obtaining the used metal. That is, the maximum permissible level of possible acquisition is 100% in theoretical justification. In practice, it is impossible to achieve such a level, but it is an ideal condition. Therefore, many research laboratories are working on alternative options for creating batteries without the involvement of rare earth metals.

The analyzed trends in the development of electric vehicles and the current technologies in the processing of rechargeable batteries, as well as the mathematical justification of recycling with an orientation towards an ideal model, allow us to predict further trends in the development of the processes of recycling EV batteries. However, we note that future scenarios are entirely conditional and depend on numerous factors that cannot be considered (new fundamental discoveries in nanotechnology and the creation of new electromagnetic and plasma technologies).

Given that the leading developed countries have decided on a complete transition from cars with internal combustion engines to electric vehicles, we highlight the following possible scenarios:

1. Full-fledged development of the recycling industry as a sphere of individual entrepreneurship, when leading specialists of well-known companies will start their own small and medium-sized battery recycling companies.
2. Further development of battery recycling based on a closed cycle logistics chain of complete recycling mobile electric batteries will work in the European and American markets.
3. Formation of innovative and breakthrough technologies based on research and experimental chemical laboratories to simplify battery recycling. Accordingly, the formation of a subsidiary in the organizational structure of a large manufacturer of electric vehicles.
4. The creation of a fundamentally new battery that will work on other fundamental physicochemical and electromagnetic principles is an evolutionary course of development that will lead to a complete transformation not only of storage batteries but also of the production of electric vehicles.

5 Discussion

Currently, battery production is constrained by the needs of EV manufacturers (Tesla, Nissan, Renault and Lifan). The manufacturers determine the principal and structural composition of these batteries. The problems with labeling complex, scientifically intensive, and chemically and physically dangerous batteries still need to be addressed by the attention of manufacturers and ecologists. Attention is paid to these problems in the context of environmental safety as a necessity and condition of general production without considering the specifics of using rare earth elements. It is necessary to work out the technology of marking and hatching batteries for complete unification and disposal needs from the point of view of environmental safety and sustainability.

Regarding the forecast of battery disposal, we predict that it will surely increase in volume, a prediction at odds with the forecasts of Alexis Gerossier, who noted that compared to 2017, there will be an increase in the use of EVs until 2023. However, by the end of 2030, their use will decline significantly.

According to the mathematical model, the disposal process can be ideal only when all components are recycled and reused in the production of batteries. However, the reality is far from ideal models due to the many complicated factors to consider. One of them was considered in this study—the marking system. Therefore, the processing technology must be adjusted for a specific manufacturer. From an economic point of view, it is energy-consuming. Therefore, the statement about using spent batteries in other industries is quite understandable, considering the life cycle of the first and second levels of application of EV batteries [56]. An analysis of the EV battery recycling and recycling market showed [57] that China is oriented towards secondary use.

The authors find that supplies of rare-earth metals to the EU are at risk and are economically important for the industry, which indicates the need to implement full-fledged recycling of batteries with the recovery of extracted metals. The global trend is to focus on secondary processing to extract 90% of metals from batteries. In contrast, the EU and the U.S. are oriented towards recycling, and Canada has focused on developing technologies for completely recycling lithium-ion batteries [58].

Therefore, the disposal problem is only being postponed for the time being, but its solution will be implemented soon. At the same time, the solution will be implemented depending on state policy, which depends on global environmental challenges. The EU and Canada pay the most attention to complete recycling.

6 Conclusions

Although the increase in environmental protection of the transport sector from the development of EVs, the disposal of EV batteries remains a significant problem, primarily due to the use of liquid-earth elements, which determine the pace of development of electric vehicle construction, current disposal processes primarily rely on existing technologies, which have organizational influence in the formation of the accompanying industry. According to theoretical rationale, the problem of processing lithium batteries will become more acute in connection with the growing demand for lithium. Therefore, with a production volume of 536.71 trillion EVs, markets will reach an equilibrium where lithium will be obtained through the complete recycling of batteries. Until then, recycling of lithium batteries will be partial and limited. EV production is realized as forming a full-fledged market with appropriate segments and is in the phase of saturation and segmentation. Factors that contribute to the development and formation of the derivative market for the disposal of car batteries: increase in the costs of valuable non-ferrous metals for the production of batteries (nickel, zinc, magnesium), as well as rare earth elements (cobalt, lithium); the system of state regulation, which prescriptively determined the nearest perspective regarding the transition of all transportation from internal combustion engines to electric motors; orientation of EV manufacturers to increase profitability by controlling the process of disposal of cars and their components to re-engage certain elements, costly ones (lithium, cobalt, copper) in repeated production.

The formation of an effective recycling market marks the establishment of appropriate chains focused on collecting, storing, sorting and recycling electric batteries. At the same time, these processes are actively manifested in East Asia since most battery production (70%) is concentrated within three companies: CATL, LG Energy Solution and Panasonic. In Europe, a characteristic feature of the EV battery recycling process is the formation of joint and subsidiary enterprises—from research chemical laboratories and local car manufacturers (Volkswagen, Volvo). Regarding the technological possibilities of disposal, the following trends were noted: there is an ongoing improvement to existing technologies based on pyrometallurgical processes and hydrometallurgical processes; there are attempts to move away from dangerous lithium technologies for the formation of cathodes and electrolytes. Possibilities for the development of innovative private businesses, as well as the formation of large regional full-cycle logistics chains for the reuse of expensive and rare earth elements, were explored.

References

1. Teixeira A, Sodré J (2018) Impacts of replacement of engine powered vehicles by electric vehicles on energy consumption and CO_2 emissions. Transp Res Part D: Transp Environ 59:375–438. https://doi.org/10.1016/j.trd.2018.01.004
2. Kester J, Noel L, Lin X, Zarazua de Rubens G, Sovacool BK (2017) The coproduction of electric mobility: selectivity, conformity and fragmentation in the sociotechnical acceptance of vehicle-to-grid (V2G) standards. J Clean Prod 207:400–410
3. Ajanovic A, Haas R (2018) Electric vehicles: solution or new problem? Environ Dev Sustain 20(Suppl 1):7–22. https://doi.org/10.1007/s10668-018-0190-3
4. Muratori M (2018) Impact of uncoordinated plug-in electric vehicle charging on residential power demand. Nat Energy 3:193–201. https://doi.org/10.1038/s41560-017-0074-z
5. Ravi SS, Aziz M (2022) Utilization of electric vehicles for vehicle-to-grid services: progress and perspectives. Energies 15:589. https://doi.org/10.3390/en15020589
6. Klingelhöfer D, Braun M, Dröge J, Fischer A, Brüggmann D, Groneberg DA (2022) Environmental and health-related research on application and production of rare earth elements under scrutiny. Global Health 18(1):86. https://doi.org/10.1186/s12992-022-00879-5
7. Prokopenko O, Järvis M, Prause G, Kara I, Kyrychenko H, Kochubei O, Prokopenko M (2022) Economic features of the use of electric vehicles in delivery services in Estonia. Int J Energy Econ Policy 12(6):340–349. https://doi.org/10.32479/ijeep.13617
8. Koval V, Sribna Y, Kaczmarzewski S, Shapovalova A, Stupnytskyi V (2021) Regulatory policy of renewable energy sources in the European national economies. Polityka Energetyczna—Energy Policy J 24(3):61–78. https://doi.org/10.33223/epj/141990
9. Glazener A, Wylie J, van Waas W, Khreis H (2022) The impacts of car-free days and events on the environment and human health. Curr Environ Health Rep 9:165–182. https://doi.org/10.1007/s40572-022-00342-y
10. Powell S, Cezar GV, Min L, Azevedo IML, Rajagopal R (2022) Charging infrastructure access and operation to reduce the grid impacts of deep electric vehicle adoption. Nat Energy 7:932–945. https://doi.org/10.1038/s41560-022-01105-7
11. Mastoi MS, Zhuang S, Munir HM, Haris M, Hassan M, Usman M, Bukhari SSH, Ro J-S (2022) An in-depth analysis of electric vehicle charging station infrastructure, policy implications, and future trends. Energy Rep 8:11504–11529. https://doi.org/10.1016/j.egyr.2022.09.011
12. Koroma MS, Costa D, Philippot M, Cardellini G, Hosen MS, Coosemans T, Messagie M (2022) Life cycle assessment of battery electric vehicles: implications of future electricity mix and different battery end-of-life management. Sci Total Environ 831:154859. https://doi.org/10.1016/j.scitotenv.2022.154859
13. Kucukvar M, Onat NC, Kutty AA, Abdella GM, Bulak ME, Ansari F, Kumbaroglu G (2022) Environmental efficiency of electric vehicles in Europe under various electricity production mix scenarios. J Clean Prod 335:130291. https://doi.org/10.1016/j.jclepro.2021.130291
14. Koengkan M, Fuinhas JA, Belucio M, Alavijeh NK, Salehnia N, Machado D, Silva V, Dehdar F (2022) The impact of battery-electric vehicles on energy consumption: a macroeconomic

evidence from 29 European countries. World Electr Veh J 13:36. https://doi.org/10.3390/wevj13020036
15. Braun A, Rid W (2017) Energy consumption of an electric and an internal combustion passenger car. A comparative case study from real world data on the Erfurt circuit in Germany. Transp Res Procedia 27(2017):468–475. https://doi.org/10.1016/j.trpro.2017.12.044
16. Athanasopoulou L, Bikas H, Stavropoulos P (2018) Comparative well-to-wheel emissions assessment of internal combustion engine and battery electric vehicles. Procedia CIRP 78:25–30. https://doi.org/10.1016/j.procir.2018.08.169
17. Kosai S, Nakanishi M, Yamasue E (2018) Vehicle energy efficiency evaluation from well-to-wheel lifecycle perspective. Transp Res Part D: Transp Environ 65:355–367. https://doi.org/10.1016/j.trd.2018.09.011
18. Pradhan S, Nayak R, Mishra S (2022) A review on the recovery of metal values from spent nickel metal hydride and lithium-ion batteries. Int J Environ Sci Technol (Tehran) 19:4537–4554. https://doi.org/10.1007/s13762-021-03356-5
19. Porvali A, Ojanen S, Wilson BP, Serna-Guerrero R, Lundström M (2020) Nickel metal hydride battery waste: mechano-hydrometallurgical experimental study on recycling aspects. J Sustain Met 6:78–90 (2020). https://doi.org/10.1007/s40831-019-00258-2ium-ion batteries. Int J Environ Sci Technol 19:4537–4554 (2022). https://doi.org/10.1007/s13762-021-03356-5
20. De Souza LLP, Lora EES, Palacio JCE, Rocha MH, Renó MLG, Venturini OJ (2018) Comparative environmental life cycle assessment of conventional vehicles with different fuel options, plug-in hybrid and electric vehicles for a sustainable transportation system in Brazil. J Clean Prod 203:444–468. https://doi.org/10.1016/j.jclepro.2018.08.236
21. Breuer JL, Samsun R, Stolten D, Peters R (2021) How to reduce the greenhouse gas emissions and air pollution caused by light and heavy duty vehicles with battery-electric, fuel cell-electric and catenary trucks. Environ Int 152. https://doi.org/10.1016/j.envint.2021.106474
22. Neugebauer M, Zebrowski A, Esmer O (2022) Cumulative emissions of CO_2 for electric and combustion cars: a case study on specific models. Energies 15:2703. https://doi.org/10.3390/en15072703
23. Islam MT, Iyer-Raniga U (2022) Lithium-ion battery recycling in the circular economy: a review. Recycling 7:33. https://doi.org/10.3390/recycling7030033
24. Islam MT, Huda N, Baumber A, Hossain R, Sahajwalla V (2022) Waste battery disposal and recycling behavior: a study on the Australian perspective. Environ Sci Pollut Res Int 29:58980–59001. https://doi.org/10.1007/s11356-022-19681-2
25. Sun S, Jin C, He W, Li G, Zhu H, Huang J (2021) Management status of waste lithium-ion batteries in China and a complete closed-circuit recycling process. Sci Total Environ 776:145913. https://doi.org/10.1016/j.scitotenv.2021.145913
26. Sommerville R, Zhu P, Rajaeifar MA, Heidrich O, Goodship V, Kendrick E (2021) A qualitative assessment of lithium ion battery recycling processes. Resour Conserv Recycl 165:105219. https://doi.org/10.1016/j.resconrec.2020.105219
27. Li L, Bian Y, Zhang X, Xue Q, Fan E, Wu F, Chen R (2018) Economical recycling process for spent lithium-ion batteries and macro- and micro-scale mechanistic study. J Power Sources 377:70–79. https://doi.org/10.1016/j.jpowsour.2017.12.006
28. Lien L (2018) Recycling lithium batteries using membrane technologies. Meet Abstr MA2018-01, p 611. https://doi.org/10.1149/ma2018-01/4/611
29. Sribna Y, Koval V, Olczak P, Bizonych D, Matuszewska D, Shtyrov O (2021) Forecasting solar generation in energy systems to accelerate the implementation of sustainable economic development. Polityka Energetyczna—Energy Policy J 24(3):5–28. https://doi.org/10.33223/epj/141095
30. Liu L, Liu S, Wu L, Zhu J, Shang G (2022) Forecasting the development trend of new energy vehicles in China by an optimized fractional discrete grey power model. J Clean Prod 372:133708. https://doi.org/10.1016/j.jclepro.2022.133708
31. Rietmann N, Hügler B, Lieven T (2020) Forecasting the trajectory of electric vehicle sales and the consequences for worldwide CO_2 emissions. J Clean Prod 261:121038. https://doi.org/10.1016/j.jclepro.2020.121038

32. Deloitte Insights (2020) Electric vehicles setting a course for 2030 https://www2.deloitte.com/content/dam/insights/us/articles/22869-electric-vehicles/DI_Electric-Vehicles.pdf
33. IEA (2022) Global EV Outlook 2022 securing supplies for an electric future. Retrieved from https://iea.blob.core.windows.net/assets/ad8fb04c-4f75-42fc-973a-6e54c8a4449a/GlobalElectricVehicleOutlook2022.pdf
34. EV-volumes (2022) The electric vehicle worlds sales database. Global EV sales for 2022 H1. Retrieved from https://www.ev-volumes.com/
35. Statista (2022) Electric vehicles—Worldwide. https://www.statista.com/outlook/mmo/electric-vehicles/worldwide
36. Geology for investors (2017) Metals and electric cars—a revolution in the making. https://www.geologyforinvestors.com/metals-electric-cars-revolution-making/
37. Trading Economics (2022) https://tradingeconomics.com/commodity/lithium
38. Investing (2022) Markets. https://www.investing.com/
39. Metals for Clean Energy (2022) Pathways to solving Europe's raw materials challenge. https://eurometaux.eu/media/jmxf2qm0/metals-for-clean-energy.pdf
40. Flyingpenguin (2022) Unsafe by design: more and more Teslas Catch on Fire. https://www.flyingpenguin.com/?p=40123
41. Visual Capitalist (2022) Land of waste: American Landfills by State. https://www.visualcapitalist.com/sp/land-of-waste-american-landfills-by-state/
42. Medium (2022) Challenges facing the battery industry in Africa & Solutions. https://medium.com/batterybits/challenges-of-the-battery-industry-in-africa-solutions-e2a17a48966a
43. TechStartups (2021) This Chinese electric car tech startup left its car lot to rot after filing for bankruptcy. https://techstartups.com/2021/05/26/chinese-electric-car-tech-startup-left-lot-rot-filing-bankruptcy/
44. Wojnowska-Baryła I, Bernat K, Zaborowska M (2022) Plastic waste degradation in landfill conditions: the problem with microplastics, and their direct and indirect environmental effects. Int J Environ Res Public Health 19:13223. https://doi.org/10.3390/ijerph192013223
45. Engadget (2021) Chinese battery maker says it's ready to produce cobalt-free EV power packs at scale. https://www.engadget.com/svolt-colbalt-free-battery-scale-production-234534595.html
46. Chevpost (2022) Tsinghua team develops super-long cycle life solid-state battery. https://cnevpost.com/2022/07/11/tsinghua-team-develops-super-long-cycle-life-solid-state-battery/
47. Seeling AlphaVeolia (2022) Why this French Environmental Services Business Could Be A Long Term Double. https://seekingalpha.com/article/4553891-veolia-why-this-french-environmental-services-business-could-be-a-long-term-double
48. Volkswagen (2021) From old to new—battery recycling in Salzgitter. Retried from: https://www.volkswagenag.com/en/news/stories/2021/01/from-old-to-new-battery-recycling-in-salzgitter.html
49. Renault Group (2019) The challenges of recycling electric car batteries. https://www.renaultgroup.com/en/news-on-air/news/the-challenges-of-recycling-electric-car-batteries/
50. Duesenfeld (2019) High recycling efficiency leads to outstanding environmental balance. https://www.duesenfeld.com/ecobalance.html
51. Umicore (2022) Battery recycling technologies and announces award with ACC. https://www.umicore.com/en/newsroom/news/new-generation-li-ion-battery-recycling-technologies-and-announces-award-with-acc/
52. Mezha (2022) 10 largest manufacturers of batteries for electric vehicles. Infographics. https://mezha.media/2022/02/03/10-vyrobnykiv-avto-akumuliatoriv/
53. London Metal Exchange (2021) LME lithium hydroxide CIF (Fastmarkets MB). https://www.lme.com/Metals/EV/LME-Lithium-Hydroxide-CIF-Fastmarkets-MB#Trading+day+summary
54. Fastmarkets (2021) Lithium hydroxide monohydrate 56.5% LiOH.H2O min, battery grade, spot price cif China, Japan & Korea. https://www.fastmarkets.com/commodity-price/lithium-hydroxide-monohydrate-56-5-lioh-h2o-min-battery-grade-spot-price-cif-china-japan-and-korea-mb-li-0033

55. Chiat J (2022) Morgan Stanley Lithium bears predict price plummet. Aust Bus Rev. https://www.theaustralian.com.au/business/stockhead/morgan-stanley-lithium-bears-predict-price-plummet/news-story/c7e0d1d805c63b9acca47a9fabb9040a
56. Haram MHSM, Lee JW, Ramasamy G, Ngu EE, Thiagarajah SP, Lee YH (2021) Feasibility of utilising second life EV batteries: applications, lifespan, economics, environmental impact, assessment, and challenges. Alex Eng J 60:4517–4536. https://doi.org/10.1016/j.aej.2021.03.021
57. Wu C, Zeng W (2022) Gel electrolyte for Li metal battery. Chem Asian J 17:e202200816. https://doi.org/10.1002/asia.202200816
58. NREL (2018) Impact of Li-ion battery recycling on the supply chain. https://www.nrel.gov/docs/fy20osti/71610.pdf

Sustainable Electrical Energy Management in the Energy Saving System Based on Analytical and Logistic Approach

Natalia Kuzmynchuk, Tetiana Kutsenko, Aida Zhagyparova, Botagoz Saparova, Azizbek Kydykov, and Serhiy Konokhov

Abstract The article is dedicated to developing analytical tools for sustainable electric energy management in the energy savings system based on the consumer-regulator model and analytical and logistic approach principles. It has been established that one of the tools for sustainable electric energy management is energy efficiency and energy savings to increase efficiency in using the energy resources available to ensure the full functioning of the domestic energy system. Involving a household consumer in sustainable electric energy management processes as a business partner will allow one to balance all market participants' interests in self-regulating the electricity consumption amount to meet their own needs. The features of new groups of household consumers of energy have been clarified and identified based on their readiness for technical participation in the electricity distribution system and load regulation, as well as their potential to attract funds to the development of the energy system. A complex model of a consumer involved in regulating electricity consumption has been proposed using an analytical and logistic approach. That proved it possible to reduce the power system load during the peak hours at 10:00 p.m. by 8.18 kWh, particularly for the «C-3» household type during the winter period, and to balance the nighttime declines in electricity consumption at the level of about 6 kW during the 11:00 p.m. to 7:00 a.m. period. It has been substantiated that consumers' transition to nighttime consumption should be accompanied by improved tariff schemes to balance the interests of both household consumers and energy suppliers based on sustainable management principles.

N. Kuzmynchuk · T. Kutsenko (✉)
V. N. Karazin Kharkiv National University, 4 Svobody Sq., Kharkiv 61022, Ukraine
e-mail: chkutsenko@gmail.com

A. Zhagyparova · B. Saparova
L.N. Gumilyov Eurasian National University, 2 Satpaeva St., Nur-Sultan 010000, Kazakhstan

A. Kydykov
Kyrgyz State Technical University Named After I. Razzakov, 66 Prospect Mira, Bishkek 720044, Kyrgyzstan

S. Konokhov
Vadym Hetman Kyiv National Economic University, 54/1 Prospect Peremogy, Kyiv 03057, Ukraine

V. Koval and P. Olczak (eds.), *Circular Economy for Renewable Energy*, Green Energy and Technology, https://doi.org/10.1007/978-3-031-30800-0_4

Keywords Energy efficiency · Energy saving · Sustainable management · Modeling · Analytical and logistic approach

1 Introduction

Existing systematic problems in the Ukrainian energy sector make it necessary to take some actions to ensure that the energy system operates sustainably. Resolving such tasks will contribute to creating and implementing the energy sector management strategy based on sustainable energy market functioning principles, compliance with energy security, and eliminating potential threats and risks that can negatively affect the energy sector's condition.

The reasons determining the need for sustainable management of the energy market in general and eclectic energy, in particular, are the unstable situation in the energy sector, the strengthening of crisis phenomena due to global instability, and the threat of complete controllability loss in the energy system. To date, about 40% of Ukrainian energy infrastructure has been damaged, translating into a $3.6 billion loss for the energy industry [1].

A significant reduction in energy capacities limits the access to electricity for household and industrial consumers, deepens the asymmetry between the volumes of production and using electric energy, exacerbates the problems of particular regions' unbalanced development, and disrupts the existing mechanisms of the electric energy market operation. Such a situation requires creating certain energy-saving mechanisms, eliminating prerequisites for a power outage, and disrupting the power system's operation. The need for sustainable electric energy management in the energy-saving system stems from the importance of creating certain management mechanisms for resolving the issues of strengthening the security of all energy system elements, including generating facilities and distribution networks that deliver electricity to end users.

2 Literature Review

The main goal of the Energy Strategy until 2050, developed with the support of the International Energy Agency (IEA), the technical assistance of the Great Britain government, and the advisory support of KPMG, is to restore the Ukrainian energy system, which has suffered significantly due to damage to critical infrastructure. The program of IEA cooperation envisages solving problems in the key areas of domestic energy industry development: security, energy conservation, transition to clean energy, and energy sector restoration.

The long-term strategy for Ukrainian energy sector formation and further development takes into account modern challenges and is aimed at ensuring the conditions for sustainable domestic economic development based on reliable, sustainable, and environmentally friendly energy sources in the context of climate neutrality. Increasing the efficiency of using limited energy resources based on energy conservation principles will determine the energy market's sustainable development in the upcoming decades. The main patterns of energy efficiency and rational electricity usage have been studied in the works of foreign and national researchers.

Prokopenko et al. [2] outlined ways to improve Ukraine's energy system based on European norms and standards and taking into account the specifics of the country. The principles of improving state support for the energy sector described in their study are based on balancing the interests of the state, energy consumers and investors. The issues of developing an effective energy-saving mechanism and ensuring the effectiveness of energy market functioning are the subject of research by Malinauskaite et al. [3] and Kovalko [4]. The research methodology of energy market functioning systems in structuring the relations between consumers included in the electricity consumption regulation system and electricity suppliers, as well as taking management decisions regarding their interaction, has been substantiated in a paper [5]. The issues of increasing the efficiency of solving energy problems with investments and R&D Costs in energy technologies are considered in the study [6].

The issue of ensuring free access to energy resources for household and industrial consumers and reducing their costs based on creating an integrated management system in the national energy sphere has been disclosed in a paper [7]. Such a system, in addition to developing and ensuring competitive relations between producers and suppliers of electric energy, includes approaches to solving the problems of regulating the activity of natural monopolies by creating a consumer market for energy products and actively involving private consumers in energy savings and efficient consumption. The multi-vector nature of resolving the energy efficiency and energy saving is-sues, which are the priority energy policy directions for the European Union countries, is outlined in studies [8, 9]. Taking energy efficiency measures is aimed at reducing household and industrial consumers' energy resources consumption volume. It creates prerequisites for preventing the decline in production and further recovery of the domestic economy, increasing its competitiveness level, and meeting the population's social needs. Developments allow one to choose how to implement state programs and regional projects in electric energy generation, distribution, and supply.

The features of the logistic approach to justifying the choice of energy-efficient measures have been studied in papers [10, 11]. In current conditions, a global energy and information complex is being formed, which creates integral mechanisms for managing energy flows that ensure supplying the maximum possible volume of energy resources to end users, provided that the cost of their consumption is minimized [10]. Kurbatova et al. [12] explored the balancing challenges of Ukraine's energy system caused by the high penetration of renewable energy and the impact of the COVID-19 pandemic on the energy sector. In modern conditions, it is still possible to implement researchers' proposals to reduce threats to the stable operation of Ukraine's energy system by installing energy storage capacities and stimulating the development of flexible renewable energy capacities.

The issues of ensuring the ecological safety of the region's population are actualized in the context of the growth increase in the natural resources, accompanied by the increase in anthropogenic pressures on the environment and humans [13]. The results of the research found in the scientific literature allow us to move on to a systematic study of the development and implementation of sustainable approaches to managing the energy-saving system in the context of global instability and the national economic crisis, as well as solving the social and economic problems of the country.

The urgency of resolving the outlined issues necessitated a detailed analysis of energy efficiency and energy-saving mechanisms in sustainable energy sector management based on an analytical and logistical approach. Such an approach should ensure energy security and achieve energy policy goals in the long term.

The study aims to develop analytical instruments for sustainable electrical energy management in the energy-saving system based on developing the consumer-regulator model to strengthen energy security in conditions of global instability and the threat of losing the controllability of the energy system, following analytic and logistic principles.

3 Research Methodology

Both general and special scientific research methods have been used to achieve the goal. Methods of generalization, comparison and deductive and inductive analysis have been used to specify the features and identify new groups of energy consumers. The idealization method allowed us to build a model of a consumer involved in an electricity consumption regulation system. Statistical analysis and graphics methods have been used to create a model of electricity consumption during one day by consumers of different types by seasons when applying basic and differential three-zone tariff, which depends on the time period of the day. Here is the mathematical representation of the complex model of a consumer involved in regulating electricity consumption:

$$P(t) = \sum_{i=1}^{n} Pi(t), \tag{1}$$

where Pi is the energy receiver capacity; n is the number of displacement steps; i is the energy receiver number.

$$Pj(t) \rightarrow Pj(t + \Delta tn), \tag{2}$$

where Pj (t) is the energy receiver; j is the number of energy receiver that is transferred; Δt is the calculated time period duration.

$$Wd = \sum_{k=0}^{23} \mathrm{P(t)} 1h, \tag{3}$$

where Wd is daily consumption; h is daily consumption per hour.

While improving the model by considering the amount of electricity consumption during different seasons of the year, it is proposed to additionally form seasonal patterns of electricity usage by household consumers and define separate subtypes. On such conditions, the extended model will allow one to study the consumption dynamics depending on the year period (season), each of the following duration: Summer—92 days; Autumn-Spring—91 days; Winter—182 days. The extended model built according to seasonal load for consumers of all subtypes also considers using a multi-zone electricity metering device.

4 Results

Experts of the International Energy Agency (IEA) claim that energy efficiency will play the most important role (40%), along with renewable energy sources (30%), in reducing carbon dioxide emissions and preventing further increase in environmental temperature by 2050 [14]. In the Energy Security Strategy of Ukraine [15], the task of implementing measures and programs to increase energy efficiency in various spheres of the national economy, particularly in the housing and communal sphere, was set as a priority direction to achieve energy efficiency in using energy resources. That is why, in order to deepen the essential understanding and carry out a more detailed analysis of the interaction between electric energy market subjects within the analytical and logistic approach, the task was set to substantiate the model of interaction between consumers involved in electricity consumption regulation and energy suppliers in the framework of Ukrainian energy market formation and market entity role, aimed at obtaining practical results to confirm the formulated hypotheses. The economic effect of involving household consumers in electricity consumption regulation systems manifests in using their financial resources. Transforming household consumers into energy system partners results in their interest in managing energy usage effectively to meet their needs [16, 17]. For this purpose, we con-sider it expedient to introduce an additional system of classification features (Table 1), which more fully reflect the changes that have taken place in consumer psychology over the last decade. It is based on such paired and mutually exclusive concepts as «technologically active ↔ technologically inert» and «investment active ↔ investment inert», which are considered as attributes (characteristics) of separate groups of household consumers.

Type «A», the most widespread consumer type, is characterized by minimal efforts to improve the existing energy consumption system and, therefore, cannot even potentially participate in regulating the energy system load. Type «B» includes consumers who make certain efforts for their microclimatic comfort but do not consider the

Table 1 New groups of household consumers of energy

Attributes (features)	Investment inert	Investment active
Technologically inert	A	B
Technologically active	–	C

interests of other energy market participants. Type «C» are consumers with technical abilities to control their own energy consumption and are willing to invest their funds in developing a sustainable energy consumption management system. The share of such consumers is gradually increasing, which characterizes them as potential business partners of energy supply companies, able to cooperate to improve the system services sector. As an element of sustainable electric energy management in the energy saving system in the context of analytical and logistic approach to electricity market improvement, approaches to modeling electricity consumption distribution by type «C» household consumer and partner during the day with a multi-zone electronic metering device are proposed.

The practical goal of the proposed model is to balance the amount of household electricity consumption based on more even load distribution, which will allow one to balance the power system's nighttime decline. On this basis, it is possible to involve a consumer as a partner and regulator in the country's energy system. A model of a consumer involved in an electricity consumption regulation system has been formed based on identifying the following three types of technologically and investment-active households, which differ in the number and capacity of household electrical appliances available. The proposed model considers a distinct difference in the number and characteristics of electrical appliances a household is equipped with, the cost of connecting a household to the distribution grid, and the income level.

That made it possible to distinguish three consumer subtypes:

«C-1» are households with a limited number of low-power electrical appliances and a low standard of living;
«C-2» are households with a basic set of electrical appliances of various capacities with an average standard of living;
«C-3» are households that use a large number of powerful electrical appliances and a high standard of living.

In turn, household appliances are divided into two groups: electrical appliances that can be used at night without any breakdown risk and electrical appliances with this option not provided. Using a well-grounded model, graphical modeling of electricity consumption by a consumer involved in regulating electricity consumption was carried out at a regular tariff depending on monthly consumption volumes and at a tariff differentiated by time periods for different seasons of the year. The obtained picture showing the dynamics of the basic demand for electricity in the studied households during different seasons is demonstrated in the example of energy consumption by «C-3» subtype household (Fig. 1).

The obtained results regarding the volumes of electricity consumption, during the winter season in particular, confirmed the hypothesis as to the need for stricter

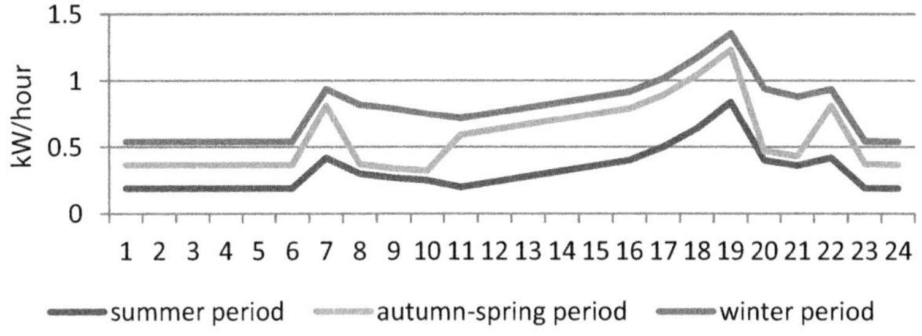

Fig. 1 The model of the basic demand for electricity by «C-3» subtype household consumers by seasons

limiting the volumes of household electricity consumption during the evening peak (Fig. 2). It made it possible to single out certain variable amount of electricity for each household subtype, which is consumed by electrical appliances that are reasonable to be connected to the power grid at night for the studied periods of the year: summer, autumn-spring, and winter.

A model for improved distribution of electricity consumption by a household consumer at night is proposed, based on the formulation of the following tasks: x_{ij} are binary variables characterizing the amount of electricity consumption by i device during j time at $x_{ij} = 1$, under condition of no consumption $x_{ij} = 0$.

According to zonal metering, the night tariff is valid from 11:00 p.m. to 7:00 a.m. [18]. The hour indicator starts at 10:00 p.m. If necessary, energy consumption may be transferred to the time adjacent to the night period, for example, from 10:00 to 11:00 p.m. Therefore, the period under consideration has 1 to 9 index and is called «adjacent nighttime consumption».

Limitation 1. The total nominal capacity of electrical appliances per hour does not exceed the set limit (one):

$$\sum_{i=1}^{n} N_i x_{ij} \leq a_j, j = 1.p \tag{4}$$

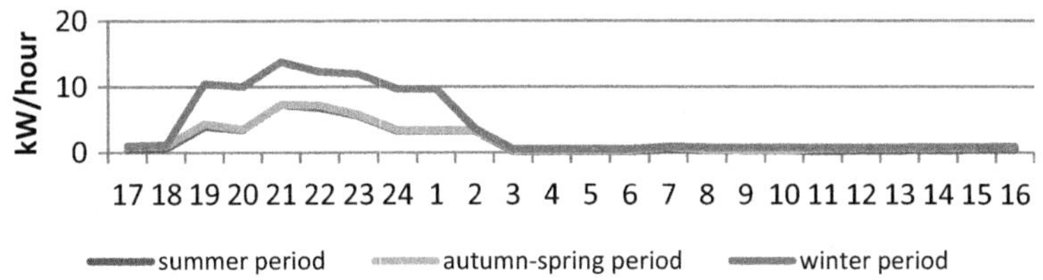

Fig. 2 The model of demand for electricity by «C-3» subtype household consumers by seasons at basic tariff

where p is the period of «nighttime adjacent consumption» (we will assume that p = 9); n is the number of electrical appliances reasonable to be connected to the power grid at night; a_j is the maximum amount of electricity consumption during j hour; N_i is the passport capacity of i household electrical appliance.

Limitation 2. The period of electricity consumption by each electrical appliance has certain duration (in hours):

$$\sum_{j=1}^{m} x_{ij} = b_i, i = 1.n \tag{5}$$

where b_i is the period of «nighttime adjacent consumption» by i household electrical appliance, respectively.

Building the objective function allows one to minimize the electricity demand. At the same time, the full night tariff is valid and transfers the household consumers to 0.4 tariffs from 11:00 p.m. to 7:00 a.m.: the night tariff effectiveness is confirmed by the shift in demand for electric energy to the period of «adjacent nighttime consumption». Analyzing additional «nighttime adjacent consumption» periods, the volume of electricity consumption is observed only under the condition of absolute loading during the 11:00 p.m. to 7:00 a.m. period, subject to compliance with the previously set limitations.

$$Z = k_2 \sum_{i=1}^{n} N_i x_{i1} + k_3 \sum_{i=1}^{n} \sum_{j=2}^{9} N_i x_{ij} \to \min, \ k_2 \gg k_3 \tag{6}$$

where k_2 is the day consumption period coefficient; k_3 is the night consumption period coefficient.

In case if technically, a specific i household electrical appliance must work continuously (certain operation modes of electrical appliances provide for their continuous operation for two or more hours), then it is necessary to add the following to the previously set limitations for the proposed mathematical model:

$$\sum_{k=1}^{10-c_j} \prod_{j=k}^{k+c_i-1} x_{ij} = 1 \tag{7}$$

where c_i is the period of continuous operation of i household electrical appliance.

This limitation has a non-linear nature, which complicates implementing a model of improved electricity consumption distribution at night for a household consumer. However, solving this problem seems possible due to including auxiliary variables in the mathematical model.

The forecast schedule of electricity consumption by household consumers during different seasons of the year when households use a basic tariff (depending on monthly consumption volumes) and a tariff differentiated by the time both for the summer period (Fig. 3) and for the autumn and spring period (Fig. 4), the following

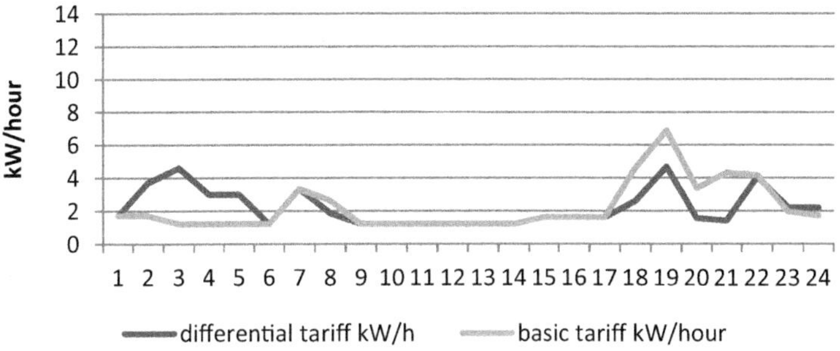

a) «C–1»

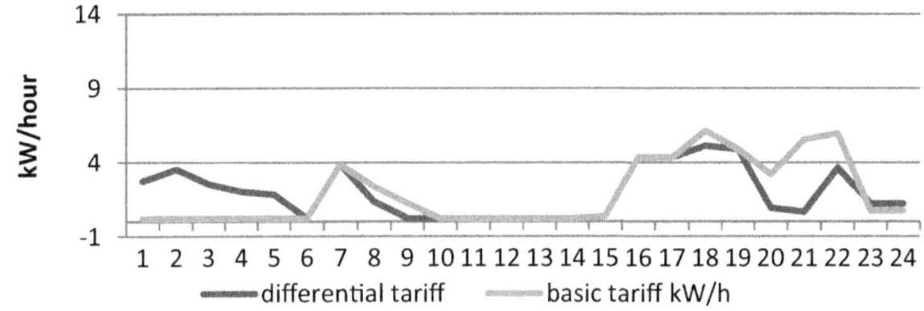

b) «C–2»

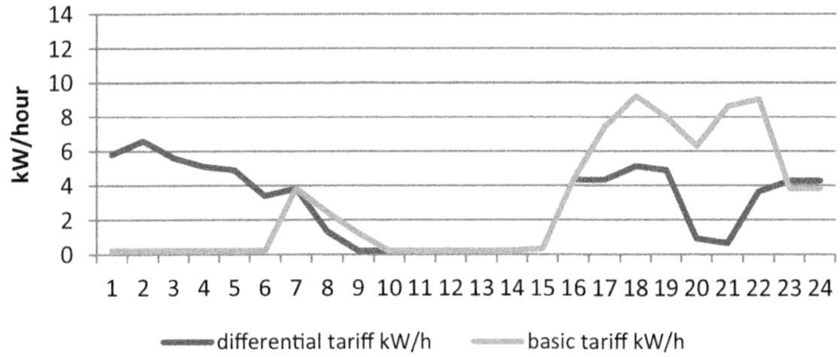

c) «C–3»

Fig. 3 The model of demand for electricity by household consumers in summer period at basic and differentiated by time periods tariffs

periods for connecting household electrical appliances are characterized, which are associated with variable demand: for household appliances, the best time of use is from 12 to 1 a.m.; for electronic gadgets, the best charging time is from 2:00 to 4:00 at night; and for an electric car, the best charging time is from 22:00 until 7:00 a.m.

For the winter period (Fig. 5), the following time to connect household electrical appliances has been calculated, attributed to variable demand: for a washing machine, the best usage time is from 1:00 to 2:00 a.m. For a boiler, the best usage time is from 2:00 to 5:00 a.m. For a dishwasher, the best usage time is from 12:00 to 1:00

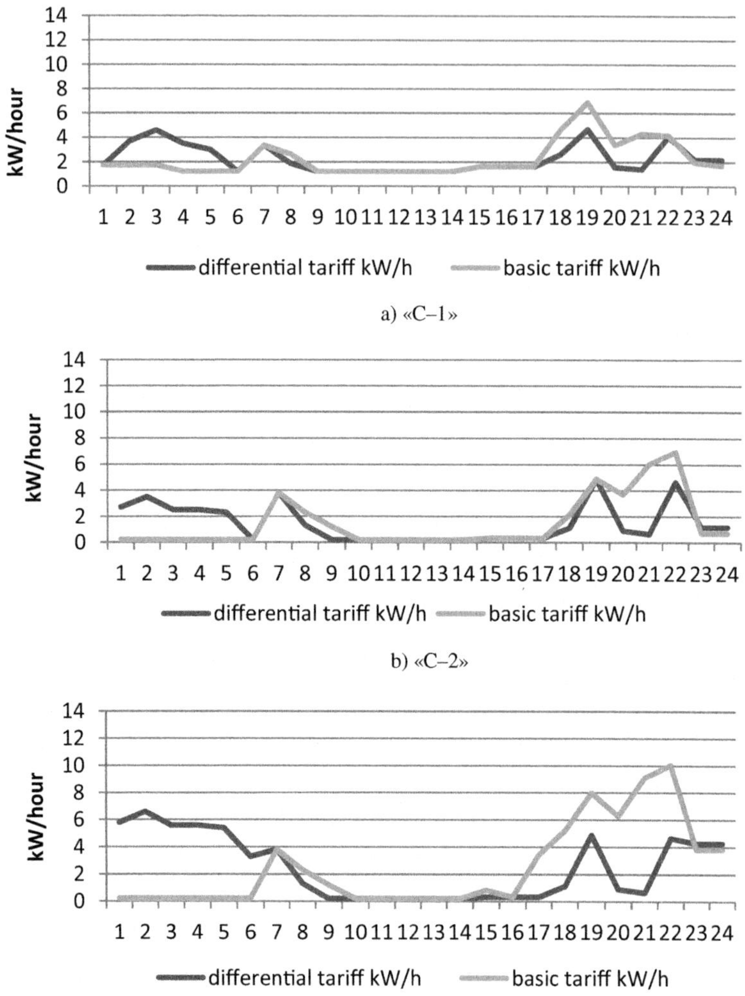

a) «C–1»

b) «C–2»

c) «C–3»

Fig. 4 The model of demand for electricity by household consumers in autumn-spring period at basic and differentiated by time periods tariffs

a.m. For smartphones and other electronic gadgets, the best charging time is from 2:00 to 4:00 a.m. For an electric car, the best charging time is from 10:00 p.m. to 7:00 a.m. For additional eclectic heaters, the best usage time is from 11:00 p.m. to 7:00 a.m. The difference between summer, autumn-spring, and the heating season is due to additional load from additional eclectic heaters.

Calculations show that a significant amount of electricity consumed by one household consumer (Fig. 5) can be shifted from the evening peak load phase to another period, particularly to nighttime during the heating season, in case consumers

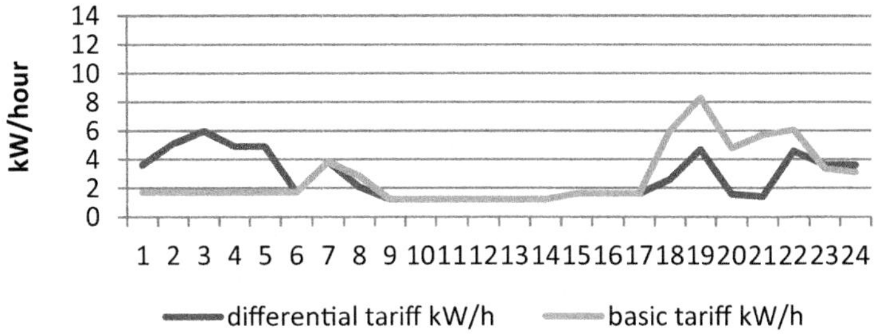

a) «C–1»

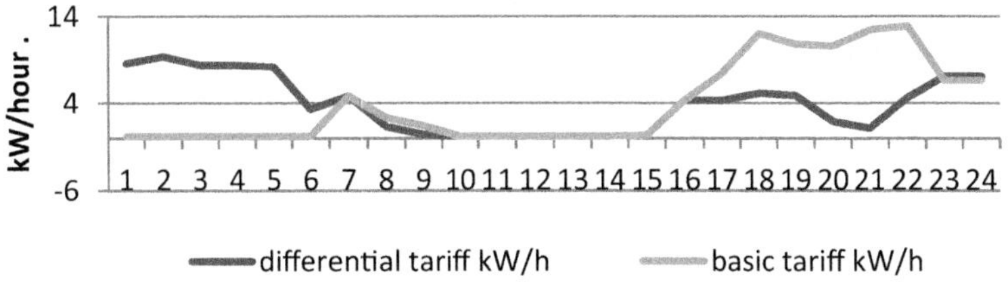

b) «C–2»

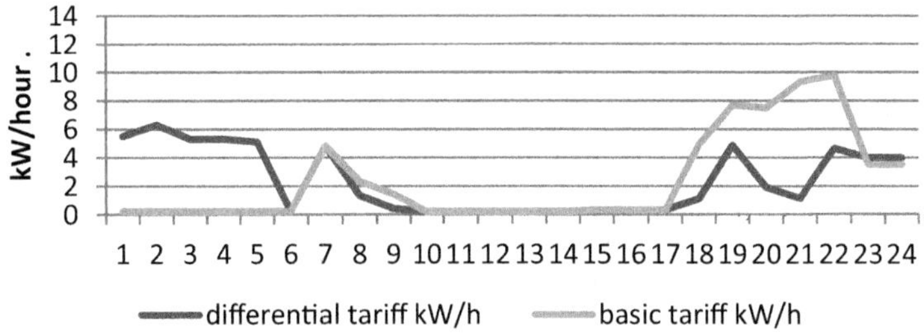

Fig. 5 The model of demand for electricity by household consumers in winter period at basic and differentiated by time period's tariffs

involved in the electricity consumption regulation system see economic interest in it.

The received calculation results showed that building the proposed housekeeping model can significantly reduce the volume of electricity consumption during peak load hours and balance nighttime declines in electricity consumption. It should be noted that the energy system will directly regulate consumption at night, applying a comprehensive approach to management system formation and controlling the electricity consumption and power system load [19]. The consumer will act as a partner and regulator, providing control over the available electrical devices in the system. The results of modeling the behavior of a consumer involved in electricity consumption regulation showed that households should be interested in switching to adjacent nighttime consumption. It is necessary to offer them favorable conditions under which a tariff with a night coefficient will stimulate consumers to night consumption without reducing the comfort level, and amount of household electrical appliances used. Developing up-to-date electricity tariffs for house-holds according to energy efficiency and saving principles will reduce consumers' financial burden when implementing sustainable electric energy management based on analytical and logistic principles.

It has to be stated that in order to resolve the issue of developing and implementing the proposed model of consumers involved in the electricity consumption regulation system, a three-zone meter is a fundamental point since sustainable eclectic energy management can be ensured based on controlling and limiting the peak load on power system only [20].

The proposed model of consumers involved in the electricity consumption regulation system, using an analytical and logistic approach, first of all, provides for dividing the households into three types of household consumers according to «C-1», «C-2» and «C-3» subcategories. The next step is to distribute these households' demand throughout the day, considering the effect of tariff differentiated by time periods, which distinguishes the following periods: peak, half-peak and nighttime. Following the goals set in the study to ensure energy efficiency and energy saving, additional parameters have been added to the housekeeping model that describes the features of electricity demand fluctuations during different periods of the year. Thirdly, we pointed out the load (variable demand) which should be transferred to the night peri-od of the differential tariff [21].

As a result, using the graphical modeling method, we received a picture of the daily load of all three «C-1», «C-2» and «C-3» types of households by household consumers' seasonal consumption features. That will, in the future, allow one to balance the interests of both consumers involved in the electricity consumption regulation system and energy suppliers, based on sustainable electric energy management principles, to ensure the energy sector security in terms of adaptation to global challenges.

5 Conclusions

Ensuring energy efficiency and savings is an essential direction of sustainable eclectic energy management and strengthening energy security if the energy system loses controllability due to a significant reduction in energy capacities. The essence of the analytical and logistic approach to modeling the consumer and regulator's behavior has been substantiated, which lies in finding the optimal balance between the interests of all market participants to supply the maximum volume of energy resources to end consumers while minimizing consumption costs. It has been proposed to involve household consumers in sustainable energy system management, giving them business partner status and allowing them to control energy use to meet their own needs independently. The classification of household consumer types has been improved according to the criteria that indicate that consumers are ready to cooperate with energy companies. That allows one to consider, on the one hand, the technical participation of consumers in the electricity distribution system and load regulation and, on the other hand, the ways of attracting their funds to develop the energy system.

There are proposed approaches to modeling the demand for electricity by type C household consumers with a three-zone meter and, therefore, the technical ability to control their energy consumption and are willing to invest their funds in developing sustainable energy consumption management systems. It has been shown that the tariff differentiated by time periods applied by a three-zone meter is beneficial for all energy system participants, profitable for consumers involved in electricity consumption regulation systems, and stimulates them to participate in sustainable energy efficiency management systems. Optimizing the electricity consumption process based on more even load distribution throughout the day will make it possible to balance the nighttime decline in the power system and attract this type of consumer to act as a partner and regulator in the country's energy system. The prospects for further research are to substantiate the approaches to developing tools for implementing energy-efficient and energy-saving ways to use and consume energy resources.

References

1. Ukraine: Economic Digest. https://www.uavarta.org/ukraina-ekonomichnyj-dajdzhest/. Last accessed 2022/12/21
2. Prokopenko O, Chechel A, Sotnyk I, Omelyanenko V, Kurbatova T, Nych T (2021) Improving state support schemes for the sustainable development of renewable energy in Ukraine [Poprawa programów wsparcia państwa w zakresie zrównoważonego rozwoju energii odnawialnej na Ukrainie]. Polityka Energetyczna 24(1):85–100. https://doi.org/10.33223/epj/134144
3. Malinauskaite J, Jouhara H, Ahmad L, Milani M, Montorsi L, Venturelli M (2019) Energy efficiency in industry: EU and national policies in Italy and the UK. Energy (Oxf) 172:255–269. https://doi.org/10.1016/j.energy.2019.01.130
4. Kovalko MP (2009) Developed energy is the basis of Ukraine's national security: analysis of trends and opportunities. Business Polygraph Printing House, Kyiv
5. Aleyn S-G (2017) Defining the concept of circular economy business model. The Carbon Trust. http://r2piproject.eu/wp-content/uploads/2017/04/Defining-theConcept-of-Circular-Economy-Business-Model.pdf. Last accessed 2022/11/15. Last accessed 2022/11/18
6. Prokopenko O, Kurbatova T, Khalilova M, Zerkal A, Prause G, Binda J, Berdiyorov T, Klapkiv Y, Sanetra-Półgrabi S, Komarnitskyi I (2023) Impact of investments and R&D costs in renewable energy technologies on companies' profitability indicators: assessment and forecast. Energies 16(3):1021. https://doi.org/10.3390/en16031021
7. Dombrovskyi O, Heletukha H (2022) Ukraine needs a new "green" energy strategy. Econ Truth. https://www.epravda.com.ua/columns/2020/01/3/655486/. Last accessed 2022/12/20
8. Pauli GA (2010) The Blue Economy: 10 years, 100 innovations, 100 million jobs. Paradigm Publications
9. Battini D (2018) Closed loop supply chain (CLSC): economics, modeling, management and control. Int J Prod Econ 183:321–329
10. Sribna Y, Koval V, Olczak P, Bizonych D, Matuszewska D, Shtyrov O (2021) Forecasting solar generation in energy systems to accelerate the implementation of sustainable economic development. Polityka Energetyczna—Energy Policy J 24:5–28. https://doi.org/10.33223/epj/141095

11. Hawks K (2006) What is reverse logistics? Reverse Logistics Mag. Winter/Spring
12. Kurbatova T, Sotnyk I, Prokopenko O, Sidortsov R, Tu Y-X (2021) Balancing Ukraine's energy system: challenges under high renewable energy penetration and the COVID-19 pandemic. E3S Web Conf 280:05007. https://doi.org/10.1051/e3sconf/202128005007
13. Kuzmynchuk N, Terovanesova O, Kutsenko T, Zyma O, Bachkir I (2021) Paradigm towards ensuring of energy saving in the crisis management conditions in the aspect of sustainable environmental development. E3S Web Conf V 255(5):01022
14. International Renewable Energy Agency (IRENA) (2019) «Towards 100% renewable energy». https://coalition.irena.org/-/media/Files/IRENA/Coalition-for-Action/IRENA_Coalition_100percentRE_2019.pdf. Last accessed 2022/12/10
15. Energy Security Strategy (2021) Decree of the Cabinet of Ministers of Ukraine dated August 4, 2021 No. 907-r https://zakon.rada.gov.ua/laws/show/907-2021-%D1%80#Text
16. Meadows DH, Randers J, Meadows DL, Behrens WW (1972) The limits to growth: a report for the club of Rome's project on the predicament of mankind. Universe Books
17. Koval V, Sribna Y, Kaczmarzewski S, Shapovalova A, Stupnytskyi V (2021) Regulatory policy of renewable energy sources in the European national economies. Polityka Energetyczna—Energy Policy J 24:61–78. https://doi.org/10.33223/epj/141990
18. Tariffs for the population. https://yasno.com.ua/b2c-tariffs. Last accessed 2022/12/06
19. Green Paper: A European strategy for sustainable, competitive and secure energy. Eur-Lex. http://eurlex.europa.eu/legal-content/EN/TXT/?uri=URISERV%3Al27062. Last accessed 2022/12/10
20. Olczak P, Matuszewska D, Lishchenko A, Zhydyk I, Koval V, Iermakova O (2022) The economic efficiency of photovoltaic energy for energy prosumers. Polityka Energetyczna—Energy Policy J 25(4):95–114. https://doi.org/10.33223/epj/155095
21. Global Engagement (2022). Marking a new era of international energy co–operation. https://www.iea.org/areas-of-work/global-engagement. Last accessed 2022/12/16

Management of Renewable Resources in the Energy Sector: Environmental, Economic and Financial Aspects

Oleksii Hutsaliuk, Nataliia Havrylova, Balsheker Alibekova, Aliya Rakayeva, Iuliia Bondar, and Yuliia Kovalenko

Abstract The purpose of the study is to substantiate the environmental, economic and financial aspects of the management of renewable resources in the energy sector with an assessment of the strategic development of renewable energy in case of Ukraine. One has assessed the potential of renewable energy sources in Ukraine, amounting to about 454.4 kWh, which is 59.2 million tons of reference fuel per year. The 2016–2020 analysis revealed that the capacity of the renewable electric power industry was increased to 10,900 MW, and the production of green electricity was up to 26 billion kW, while the generation of thermal energy from renewable resources increased to 5.85 million tons. The development of renewable resources was ensured by issuing Green and Sustainability-linked bonds for five years, amounting to 825 million USD with a yield rate of 6.785%. Directions were identified to increase the production of renewable resources in the energy sector and increase their share to 20% by the Energy Strategy of Ukraine until 2035. Therefore, there will be a demand for green and environmental investments and green bonds, indicating the placement of funds in energy-efficient and environmental projects.

Keywords Renewable energy sources · Economic and financial efficiency · Green electricity · Solar power plants · Energy sector

O. Hutsaliuk (✉)
Private Higher Education Institution, «Rauf Ablyazov East European University», 16, Nechuia-Levytskoho Str., Cherkasy 18036, Ukraine
e-mail: alex-g.88@ukr.net

N. Havrylova
Central Ukrainian National Technical University, 8, University Ave., Kropyvnytskyi 25000, Ukraine

B. Alibekova · A. Rakayeva
L.N. Gumilyov Eurasian National University, 2 Satpaeva St., Nur-Sultan 010000, Kazakhstan

I. Bondar
Flight Academy of the National Aviation University, 1, Stepana Chobanu Str., Kropyvnytskyi 25005, Ukraine

Y. Kovalenko
State Tax University, 31 Universytetska Str., Irpin 08200, Ukraine

V. Koval and P. Olczak (eds.), *Circular Economy for Renewable Energy*, Green Energy and Technology, https://doi.org/10.1007/978-3-031-30800-0_5

1 Introduction

Renewable resources are an essential component of the energy sector, the impact of which is increasing on a national and global scale. Energy resources that have traditionally been used are gradually being exhausted, and renewable ones, such as solar energy, are gaining more and more popularity and relevance due to the expansion of their availability and state support.

The economically justified potential of renewable energy sources in Ukraine is about 454.4 kWh, which is equal to 59.2 million tons of reference fuel per year [1]. Having joined the European Energy Community, Ukraine must produce 25% of its electricity from renewable energy sources by 2035. By the Decision of the Cabinet of Ministers of Ukraine «On the implementation of Directive 2009/28/EC and amendments to Article 20 of the Agreement on the Establishment of Energy Cooperation», according to which national goals concerning renewable resources in the energy sector are defined in order to provide guarantees to investors and carry out incentive measures for the development of the latest and innovative technologies [2, 3].

According to this document, Ukraine was obliged to reach 11% of total renewable energy in final consumption by 2020, which is a powerful lever in producing renewable energy resources. The measures specified in this document contribute to the implementation of the state policy on taking measures for the management of renewable resources in energy, which will allow significantly, reduce the consumption of traditional fuel and energy resources for consumption by economic sectors and, therefore, will strengthen the level of energy security of the country.

Considering the global problems with energy resources and the increasing dependence of the national economy on imported energy sources, the need to find new sources of energy increases and private businesses, foreign investors, and the state attract new sources of financing. Considering the harmfulness of many sources of energy, namely gasoline, diesel and nuclear fuel, on the ecosystem, there is an increasing need to find new renewable sources of energy, namely wind and solar power, which are the most affordable and safe. Therefore, studying renewable energy sources' environmental, economic and financial components is relevant and requires detailed research.

2 Literature Review

The growing importance of renewable resources in energy determines the development of these studies. Herneho and Liakhova [2] paid attention to financing the development potential of alternative energy in Ukraine. Shulha and Ivashchenko [4] actively investigated the foreign experience of implementing, developing and expanding renewable energy areas. Skorokhod and Horbach [5] studied problems of the green economy in the European Union countries.

The research by Lishchuk et al. [6] is dedicated to the study of strategic priorities and ways of developing renewable resources in energy. Economic mechanisms of the development of renewable resources in the energy industry are revealed in the scientific research of Riazanova [7], focused on scientific principles of the organizational and economic mechanism of the development of renewable resources in energy. Prokopenko et al. [8] thoroughly analyzed the state of renewable energy in Ukraine and its legislative support. Based on the analysis, the researchers formulated recommendations for improving state support schemes for the country's sustainable development of renewable energy.

Twidell [9] examines renewable energy sources from the perspective of basic scientific theory and then the application of environmental impact and socio-economic aspects. Kurbatova et al. [10] analyzed the trends in the development of renewable energy, the dynamics and structure of consumption and export of electricity in the pre-epidemic and epidemic periods, and also identified the main challenges to the operational security of the Unified Energy System of Ukraine.

Prospects for the use of renewable energy sources (wind, solar, waves and biomass) in the development of sustainable development strategies are explored by Lund [11]. Such strategies usually involve three main technological changes: energy conservation on the demand side, more efficient energy production and replacing fossil fuels with various renewable energy sources. Researchers [12–14] also pay attention to the development of startups and enterprises in the energy industry, including their social responsibility.

Sotnyk et al. [15] substantiate the role of electric transport in strengthening energy security by switching to renewable energy technologies and reducing the use of fossil fuels.

The global association of scientists of the Green Economics Institute (UK) also developed the modern interpretation of the theoretical foundations of the green economy. Kennet [16], the Director of this Institute, believes that the green economy returns science from exclusively quantitative dimensions of economic and human beings and expands the boundaries of the mainstream using complex and interdisciplinary methods and tools.

The UN and the OECD consider the green economy as the further transformation and clarification of the concept of sustainable development based on the principles of interaction between society and nature and involves harmonizing economic, social development and environmental protection [17].

Based on the example of Denmark, this paper discusses the challenges and prospects of converting existing energy systems to 100% renewable energy systems. Conclusion: such a scenario is possible. The necessary renewable energy sources are available, and further technological advancement in the energy system can help create a renewable energy system. Transport sector transformation technologies and the introduction of flexible energy system technologies are significant. Therefore, large-scale renewable energy implementation plans should include strategies for integrating renewable sources into coherent energy systems influenced by energy conservation and efficiency measures.

In Ukraine, areas of the strategic development of renewable energy resources, advantages and threats of the national energy sector, and promotion of the increased energy efficiency of society are considered in the Energy Strategy of Ukraine until 2035 [18]. Considering numerous theoretical studies of leading Ukrainian scientists and the legislative basis, the problem of managing renewable resources in the energy industry, especially its ecological, economic and financial aspects, remains unresolved.

The purpose of the study is to evaluate theoretical and practical aspects of renewable resource management in energy, to identify problematic aspects of the ecological, technological, production and financial state, as well as to consider the strategic development of renewable solar energy under the standards of sustainable project management.

3 Methodology

At the current stage, the assessment of ecological, economic and financial efficiency of resource conservation at the level of business entities is carried out using the following approaches:

- Determination of ecological, economic and financial efficiency of resource-saving measures is carried out based on changes in total resource capacity.
- Calculation of ecological, economic and financial efficiency of resource conservation based on indicators of economic efficiency of investment projects and resource intensity.

Resource-saving projects are evaluated using indicators of net present value, internal rate of return and payback period. The net cash flow is calculated by comparing the total energy intensity of a production unit by types of energy resources before and after the implementation of energy saving, taking into account the volume of the saved energy resource and its unit prices according to formula 1 [19]:

$$R_o = \sum_{i=1}^{m} (e_{oi} - e_{li}) O_l \mathrm{P_i} \tag{1}$$

where

R_o is the main result from the implementation of green investments, USD;

m is the number of types of energy resources, the consumption of which changes depending on the implementation of the measure;

e_{oi}, e_{li} are energy intensity of 1 dollar of manufactured products by the i type of resource before and after the implementation of the measure, respectively, unit/dollars;

O_l is the volume of production after the implementation of the measure, USD;

P_i is a unit price of the i type of energy resource, USD.

- calculation of the ecological, economic and financial effect of resource conservation. A standard indicator of the assessment of resource-saving measures is the economic effect obtained as the difference between economic results and costs from implementing projects related to implementing measures to restore resources in the energy sector. At the same time, when calculating the economic effect, ecological and social components of resource conservation are almost not considered in national practice.

However, there is a technique for evaluating the social and ecological components of the economic effect of resource conservation on business entities [20]. The annual ecological, economic and financial effect of business entities (Ee.e.) obtained from the implementation of a set of resource-saving measures is determined by the formula 2:

$$\text{Ee.e.} = \text{Er} + \text{Eep} + \text{Ecc.} - \Delta\text{A}\sum +\text{El} + \text{Elr}, \tag{2}$$

where

Er	is the annual saving of resources, USD;
Eep	is the decrease in the amount of environmental payments of business entities as a result of the implementation of resource-saving measures, USD;
Ecc	is the saving of other current costs (purchase of auxiliary materials, energy, reduction of salary costs, as labor productivity increases), USD;
$\Delta A\sum$	is the total increase in the amount of amortization deductions for resource-saving measures, USD;
El	is a part of the averted economic loss as a result of the implementation of resource-saving measures by business entities, but not included in its environmental payments, USD;
Elr	is the reduction of annual ecological, economic and financial losses of economic entities as a result of man-made emergencies in production related to the use of resources, USD.

The maximum value of the annual ecological, economic and financial effect is a criterion for making a management decision on implementing a resource-saving project.

A comprehensive approach to resource conservation at business entities includes several measures to increase the resource efficiency of production depending on periods. It requires considering the time factor when calculating ecological, economic and financial efficiency. The maximum value of the complex ecological, economic and financial effect (E_k^{ee}) is a criterion for choosing the best possible resource-saving measures implemented at business entities for a long time and is determined by formula 3 [21]:

$$E_k^{e.e.} = \sum_{i=1}^{T} \frac{E_i^{e.e.}}{(1+r)^i} \to \max \tag{3}$$

where

- i is the year of obtaining the ecological, economic and financial effect from the implementation of the set of resource-saving measures;
- T is the period of obtaining ecological and economic effects of resource conservation, years;
- $E_i^{e.e}$ is the ecological, economic and financial effect of the implementation of the set of resource-saving measures, which is received by economic entities in the i year;
- r is the standard of bringing different time costs.

Taking into account formulas (2) and (3), the complex ecological, economic and financial effect can be determined by the formula:

$$E_k^{e.e.} = \frac{Er + \text{Ecc} + \text{Eepi} + \text{El} + \Delta\text{Li} - \text{Kdi}}{(1+r)^i} \to \max \tag{4}$$

where

- ΔLi is the increase in the liquidation value of the equipment in the i year, USD
- Kdi is additional capital investments in resource-saving measures in that year, USD.

When the business entity chooses the best possible resource-saving measures, it faces a situation where the holding of complex ecological, economic and financial effects comparable in size is achieved with different amounts of invested funds. The best possible choice criterion is the coefficient of ecological, economic and financial efficiency of the resource-saving measure ($k^{e.e.}$), which shows the amount of additional income per unit of capital investment during the implementation of the resource-saving measure ($k^{e.e.}$). It is determined by formula 5 [22]:

$$k^{e.e.} = \frac{E_k^{e.e.}}{\sum_{i=1}^{T}(K_{d.i}/(1+r)^i} \to \max \tag{5}$$

Resource-saving measures that have the most significant value (ke.e.) provide business entities with the most significant increase in income. They are considered the most efficient.

4 Results

The basis of renewable resources in energy is the production of an energy product, which should be interpreted as the expenditure of various types of energy used for a tangible or intangible product. Depending on the sources of origin and the degree of processing of energy products, one distinguishes primary energy (a natural resource obtained by the extraction of solar energy, coal, oil and uranium) that has not yet undergone processing and transformation and secondary energy resources that have undergone processing and formation through technological aggregates, as well as renewable energy sources, namely solar energy, water energy, geothermal energy.

The National Action Plan on the management of renewable energy resources allowed solving the following vital tasks [23]:

- the stipulated renewable electricity capacity was increased to 10,900 MW, and the production of green electricity was increased to 26 billion kW;
- thermal energy produced from renewable resources was increased to 5.85 million tons;
- the amount of energy sources from renewable resources used in the transport sector was increased to 505 thousand tons.

The results of the promotion and use of energy produced from renewable resources are shown in Table 1.

Studies show that in 2021, the specific weight of electricity obtained from renewable resources was 8.1% or 12.8 TWh; 56% of this volume is solar energy, 33% is wind energy, 8% is biomass and biogas burning, and 3% is small hydropower. In 2021, 12,804 million kWh of clean electricity were obtained from all types of renewable energy resources, which is 1941.9 million kWh, or 17.8%, more than 2020. Solar power plants produced 7670 million kWh, which is 4.8% of the total amount of electricity and 1065.4 million kWh more than in 2020 [24].

2021 was a decisive year in the development of renewable resources in the energy sector of Ukraine, as the daily production of electricity from renewable resources in the energy sector exceeded the level of generation by thermal power plants for the first time and amounted to 79 million kWh versus 77 million kWh [25]. Implementing promising projects related to renewable energy resources in Ukraine made it possible to reduce CO_2 emissions into the atmosphere. In 2021, they were reduced by almost 10.3 million tons, approximately 2.2 million cars. Private households' wind and solar power stations clearly demonstrate the positive development dynamics.

Thus, between 2018 and 2022, the total capacity of solar energy production mini-stations increased more than five times and was equal to 1205 MW at the beginning of 2022, which is almost 45 thousand units [23, 25]. A strong incentive for installing solar power plants in private households is the green rate that has significantly contributed to developing renewable energy sources. On the other hand, it has revealed many problems. Thus, the fixed green rate stimulates private households

Table 1 Results of the promotion and use of energy produced from renewable resources 2016–2020 (%)

Areas of use	2016	2017	2018	2019	2020
In heating and cooling systems	6.2	7.56	8.0	9.03	9.28
In power engineering	7.91	8.64	8.9	10.89	13.92
In the transport sector	2.1	2.44	2.2	3.07	2.47
The total share of renewable resources	5.85	6.67	7	8.08	9.19

Source Based on [1]

not only for their energy independence but also for the purposeful sale of electrical energy because of the high level of the green rate, namely 18.09 euro cents for 1 kW of solar generation and 16.26 euro cents for 1 kW of wind generation [23].

Besides, the identified serious problems include technical obstacles and direct abuses in constructing solar power plants for private households, but generally without their consumption. It is not the provision of one's energy independence but the deployment of business [23]. A significant impetus to developing renewable resources in the energy sector was the issuance of Green and Sustainability-linked bonds for five years, amounting to 825 million USD with a yield rate of 6.785%. Ukrenergo NPC became the issuer of Eurobonds. Issued bonds are with an unconditional and irrevocable state guarantee. From December 2021, Green and Sustainability-linked bonds were admitted to circulation in Ukraine by the National Securities and Stock Market Commission, which significantly expanded investment opportunities for Ukrainians. Before that, only international investors could become owners of Eurobonds [24]. Currently, 87.8 MW of wind power plants and 407.9 MW of SPPs are located in the temporarily occupied territory of Crimea, and 138 MW of wind power plants are located in the temporarily occupied territory of Donbas.

In total, 633.7 MW of renewable energy resources are in the territory occupied by Russia [23]. In Ukraine, the production of renewable energy resources is financed from different sources and different forms of ownership. Figure 1 shows forms of ownership of energy sector enterprises and sources of their financing.

Therefore, renewable energy enterprises can be established in the public and private sectors. In Ukraine, state support is often provided to enterprises receiving energy from river resources, i.e., hydroelectric power plants. At the same time,

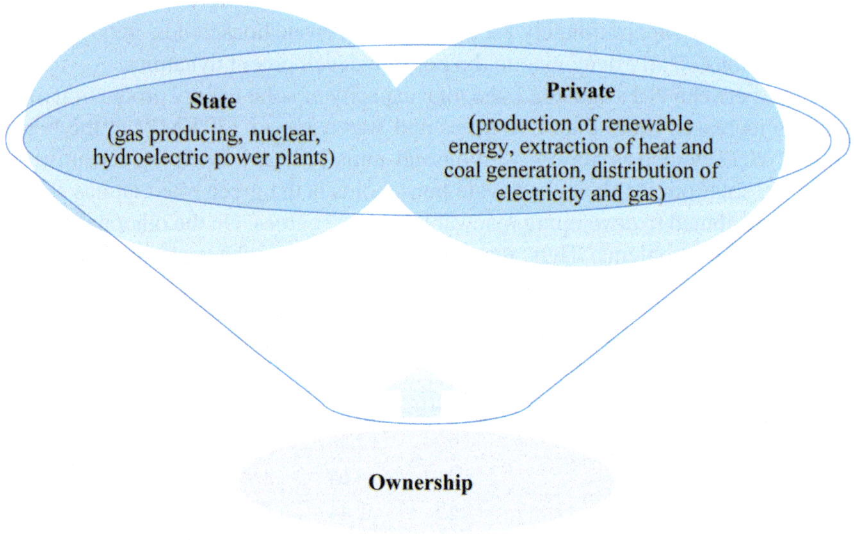

Fig. 1 Form of ownership of energy sector enterprises in Ukraine

several mechanisms aimed at supporting renewable energy have been developed and implemented in Ukraine.

Measures that stimulate the production of renewable energy sources using the green rate, preferential taxation and preferential connection to the power grid have been widely used. In Ukraine, only approximately 10% of the total production of energy resources is covered by alternative energy sources.

Ukraine is actively implementing actions to obtain green energy in the Green Deal that stipulates the increase in the production of renewable resources in the energy sector and the increase in their share to 20% by the Energy Strategy of Ukraine until 2035 [18]. Implementation of these measures will be possible due to the further implementation of projects on renewable energy resources.

The analysis of capacity and energy output by supply sources in Ukraine for 2018–2020 is shown in Table 2.

Analyzing the data in Table 2, we can see that in 2019, compared to 2018, the amount of installed electrical capacity and the output of electrical energy decreased. We can see a significant decrease in the volume of electrical energy sales by – 6699 kWh in 2020 compared to 2019. A similar situation was observed in 2020 compared to 2019. Thus, in thermal power plants, there was a decrease in the installed electric capacities in 2020 compared to 2018 and 2019 and in the volume of electricity supply for the same period. No new electrical capacities were installed at nuclear power plants during the studied period.

Analyzing capacity and energy output by supply sources in Ukraine for 2018–2020, we can see a positive trend in producing renewable energy resources. Thus, in 2019, compared to 2018, the installed electric capacity of wind power plants increased from 476 thousand kW to 795 thousand kW and to 1110 kW in 2020. At the same time, there was a noticeable increase in electricity sales volume: from 1182 million kWh in 2018 to 1760 million kWh in 2019 and 3271 million kWh in 2020. This positive trend is also observed in the operation of solar power plants. Thus, in 2018, the electric power of solar power plants was set at 1201 thousand kW; in 2019—1953 thousand kW; and in 2020—5194 thousand kW. Along with the increased installed electric capacity, there was a noticeable increase in the volume of electric energy output. Thus, in 2018, the electricity output volume was 1103 million kWh; in 2019—1883 million kWh; in 2020—5684 million kWh.

These positive changes indicate the strengthening of renewable resources' role in energy, especially solar energy. Analytical solutions of the Wärtsilä [27] company show that we can expect a gradual transition from a combined energy system to a renewable one by 2050, which can amount to 131 GW in the future. The construction of a large-scale solar power plant is planned for 2026. Its launch will be accompanied by a gradual decrease in variable costs and a decrease in operational costs [27].

Currently, the processes of developing renewable energy resources are revitalizing in Ukraine. Their financing sources are expanding, particularly at the expense of the state and private banking sectors. Thus, Oschadbank JSC, together with the international financial sector, is implementing the program: Energy efficiency in the housing sector of Ukraine. Ukrgasbank JSC, together with the European Bank for Reconstruction and Development, is implementing measures to support national

Table 2 *Analysis of capacity and energy output by supply sources in Ukraine for 2018–2020*

	2018		2019		2020		Deviation, the release of electricity kWh 2020/2019 (±)
	Installed electric power, thousands of kW	The volume of electricity supply, million kWh	Installed electric power, thousands of kW	The volume of electricity supply, million kWh	Installed electric power, thousands of kW	The volume of electricity supply, million kWh	
Thermal power plants	23,906	43,773	22,265	40,910	22,311	36,300	−4610
Thermal electric power plants	5470	10,922	5855	10,738	5890	12,837	20,099
Nuclear stations	13,835	79,383	13,835	77,948	13,835	71,249	−6699
Wind power plants	476	1182	795	1760	1110	3271	1511
Solar power plants	1201	1103	1953	1883	5194	5684	3801
Hydroelectric power stations	6242	11,826	6326	7712	6335	7415	−297
Other generating plants	378	135	415	262	463	441	179
Total	51,508	148,324	51,444	141,213	55,138	137,197	−4016

Source Based on [26]

businesses aimed at revitalizing the implementation of investment projects on the development of renewable resources in the energy sector and the increase in their energy efficiency. Raiffeisen Bank Aval JSC directed its activities to expand relations with clients joining the UKEEP international program for increasing energy efficiency and strengthening the potential of renewable resources in the energy sector. It allows banks not only to revive the level of banking services but also to stimulate the development of alternative energy in Ukraine and worldwide due to their close interaction with EPC contractors (Engineering, procurement and construction), which are specialized institutions promoting consulting services and services for the construction of energy infrastructure facilities, their commissioning, maintenance. EPC contractors, taking into account the experience of conducting renewable energy business, will increase the efficiency of management and implementation of renewable energy resources projects using their own experience and material facilities. Given the potential and business reputation of EPC contractors, this will minimize banking risks and technological risks to customers and investors.

Implementing the national strategy and focusing on the European Green Deal will allow Ukrainian businesses to increase energy saving and environmental responsibility. Therefore, there will be a demand for green and environmental investments and green bonds, indicating the placement of funds in energy-efficient and ecological projects. Foreign experience shows that green bonds are aimed at energy-saving and energy-efficient projects counteracting negative natural phenomena and climate change. Data from the European Energy Community indicate that Ukraine has joined this organization.

Renewable resources in energy are periodic or constant flows of energy that spread in nature and are limited only by the stability of the Earth. It includes solar radiation, wind, hydropower, and natural thermal energy.

The development of renewable energy has a powerful impact on the development of society because energy minerals are an exhaustible limited resource. Thus, society should produce and use an energy resource that has an unlimited available source and is permanent, not temporary. Solar energy is a permanent renewable energy resource and can be converted into energy through thermodynamic and photovoltaic pathways. The thermodynamic method converts solar energy into electrical energy using traditional schemes in thermal installations where a stream of concentrated solar radiation replaces the increase in temperature from fuel combustion.

One distinguishes the following types of solar thermal power plants: a tower type due to the concentration of solar radiation from flat mirrors-heliostats; a parabolic (tray) type where vacuum receivers-tubes with coolant are placed through parabolic cylindrical concentrators; a dish type where a solar energy receiver with a working fluid is located through the focus of a dish-shaped parabolic mirror.

Solar photo energy is a direct conversion of solar radiation into electrical energy. A photoelectric converter operates through the internal photo effect. Ukraine's geographical location and climate allow the active production of solar energy, a renewable energy resource. The average annual amount of energy provided by solar radiation in the territory of Ukraine varies annually from 1070 to 1400 kWh depending on the region. Photovoltaic equipment can efficiently generate solar energy for almost a year, especially in the southern and central regions, and for almost seven months in the northern regions. According to the results of various calculations, the feasibility of installing solar energy is approximately 4 GW. Ukraine's existing reserves of raw materials, scientific and technical facilities, and industrial capacities for the production of photovoltaic devices are capable of fully meeting the needs of local consumers and certain export needs [1].

In addition, a draft Plan for the Recovery of Ukraine has been developed, which defines the strategic priorities of the recovery and modernization of the entire energy sector, taking into account already existing international obligations and new opportunities and needs of the Ukrainian and EU energy market, arising during active hostilities, as well as determines national priorities for the restoration and modernization of the national energy sector [23]. Even in February 2022, Europe's Ukrainian energy industry was quite powerful. Ukraine has one of the highest shares of carbon–neutral generation in Europe; almost 70% of Ukrainian electricity is produced at the expense of nuclear, hydropower and renewable energy resources [25]. The operational capacity of power generation facilities as of August 1, 2022, was as follows (Fig. 2).

When considering the issue of managing renewable resources in energy, it is appropriate to consider the advantages and disadvantages of solar energy, which are presented in Fig. 3.

The given data indicate that about 4% of the generating capacity was destroyed during hostilities, and another 35% of the capacity is located in the occupied territories. In general, almost 50% of thermal generation, 30% of solar generation, and more than 90% of wind generation have been destroyed or are located in the occupied territories. Considering the advantages and disadvantages of renewable solar energy, we can determine the strengths and weaknesses of renewable solar resources in energy. The strategic development and management of renewable energy resources

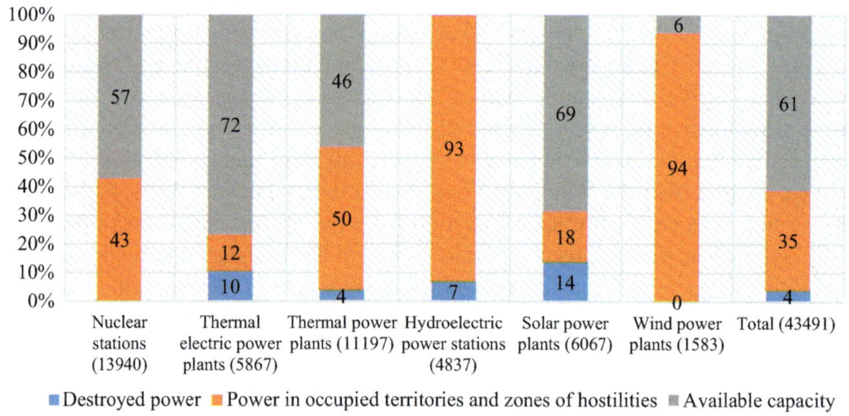

Fig. 2 Distribution of operating capacity of power generation facilities as of August 1, 2022, MW in Ukraine [25]

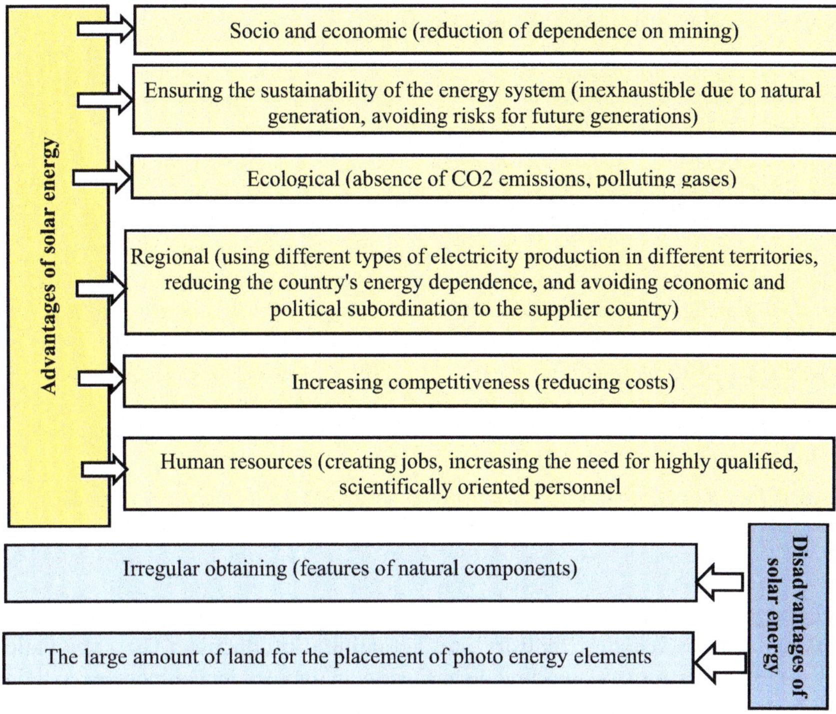

Fig. 3 Advantages and disadvantages of renewable solar energy

take into account the state's economic and financial potential, technological development, and natural and climatic possibilities. The natural and climatic conditions of Ukraine ensure the practical implementation of the development and management of renewable energy resources due to the number of sunny days providing significant generation of renewable energy sources. The production of renewable energy sources and their consumption in Ukraine is accompanied by unevenness.

Thus, maximum consumption occurs in the winter with a short global day and the maximum production in the summer with an extended global day. The constant expansion of the capacities of renewable resources in the energy sector requires an increase in investment needs. It will require additional operational costs, new balancing capacities and transformations of the energy sector's existing technical components of renewable resources. In addition, we see the increasing need to balance renewable resources in the energy sector and the maneuverability of the power system due to peak energy receipts and consumption. The rational use of strengths and improvement of weaknesses reveals the unlimited development of renewable resources in energy. It will allow meeting the need for energy carriers in case of high technological development, but the primary intellectual, financial and technological investments will be high. Table 3 shows the strengths and weaknesses of using renewable solar resources in energy.

Table 3 Strengths and weaknesses of using renewable solar resources in energy

Strengths	Weak sides
It is possible to create a station of different capacities thanks to modular designs	Irregularity of generation over time
Favorable geographical location, availability of high generation potential	Low level of national technological solutions
The possibility of attracting foreign investments	High initial costs for the creation of the accompanying infrastructure
Implementation of the world's leading technologies for the production and transformation of renewable electrical energy	Imperfect energy storage systems, low technological level of storage batteries
Production of eco-friendly energy	The weakness of the existing outdated power grid
Solving environmental, social and financial problems	Financial and political instability
Strengthening the energy security of the owner of solar panels, the region, the country	Variability of production of renewable resources in energy

Energy consumption is related to all human economic activity: heating houses, cooking, driving vehicles, industry and agricultural production. The exploitation of various energy reserves worldwide has led to an unprecedented increase in living standards [23, 24]. Today, people are so dependent on energy that it is hard to imagine how they would live without it.

Energy production significantly affects the environment. Spills of oil and oil products during extraction and transportation can destroy all living things in a huge area (water). The release of sulfur, carbon dioxide and carbon monoxide gases, nitrogen oxides, dust, soot and other pollutants accompanies the combustion of solid and liquid fossil fuels. Surface coal mining and peat mining cause changes and sometimes damage to the natural landscape.

The development of infrastructure required to extract coal, oil and gas can have a detrimental effect on landscapes, flora and fauna. Nuclear power is potentially dangerous because of potential accidents at power plants and the release of radioactive materials into the environment. Nuclear waste remains dangerous for hundreds and thousands of years. Over recent years, politicians and society have been expressing concern about the worsening of global environmental problems, such as acid rain and climate change, and assessing the impact of these processes on the environment. The topic is especially relevant for Ukraine, which has suffered from the consequences of the Chornobyl nuclear power plant accident. Although it is possible to generate energy more sustainably using renewable sources (solar, wind, hot water, wood and agricultural waste), we should admit that there is no way to generate entirely environmentally friendly energy (Fig. 4).

The management of renewable energy resources and the gradual transition to the green economy requires the following measures: development and participation in programs to increase energy efficiency and develop renewable resources in alternative

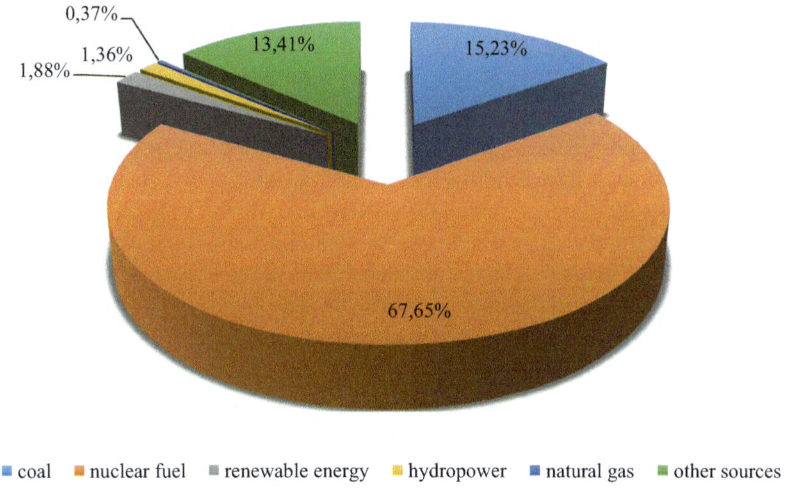

Fig. 4 Share of energy sources used in the overall structure of the 2021 balance of total electrical energy in Ukraine [26]

energy; implementation of resource-saving measures and eco-friendly production; carrying out measures to stimulate investment programs for the introduction of green businesses; expansion of environmental safety measures by introducing environmental innovations; promoting the development of ecological services, production of ecological products.

The national concept of Ukraine's environmental policy [23] defines that renewable energy sources, including solar energy, are safe because they emit fewer pollutants, lower noise and electromagnetic pollution, and reduce the amount of emissions and discharges of harmful substances. The maintenance of solar panels is carried out using simple and harmless procedures (washing, mechanical repair); the operation process does not affect the atmosphere because energy is produced physically without harmful emissions, and solar panels are made of materials that do not harm the environment. The main environmental problem of renewable solar energy is the disposal of equipment—used, spoiled, damaged solar panels [27].

According to the Ukrainian Institute of Renewable Energy specialists, the problem of disposal of solar panels was not relevant until 2030. However, military actions partially destroyed the objects of renewable solar energy, so the issue of their disposal will arise after the de-occupation of the territories. All elements of solar panels, namely glass, aluminum, plastic and thin silicon wafers, can be completely recycled and reused. Many European countries, which pay much attention to environmental protection and worry about their current and future citizens, do not find arguments against using solar electric panels. Farms in highly developed countries often use dual farming, solar farming, or agrovoltaics, which allows using the field simultaneously for solar panels and agriculture. Beans, coriander, tomato, lettuce, and greens are usually grown under solar panels. The main disadvantage here is manual processing or very small and maneuverable equipment. The advantage is the fact that dual farming can improve the situation of pollinating insects by improving natural vegetation.

However, it is worth noting the negative consequences of using solar energy. They primarily include problems related to land use and water use. The construction of an industrial solar energy base requires a large area, leading to the clearing and sorting of land (natural water bodies, mountainous terrain), which can change the structure of the soil, erosion and change the structure of water resources. However, the rational placement of solar panels (roofs, reclaimed soil) can prevent this problem. Therefore, the main ecological task of introducing and using renewable resources in solar energy is rationality in the use of areas, capacities and operational capabilities [28].

In managing renewable resources in energy, it is increasingly worth paying attention to such issues as sustainable development, global warming, ethics, social responsibility and the supply system [29, 30]. Taking into account the above, we offer our model of management of renewable resources in the energy industry, shown in Fig. 5.

The model of management of renewable resources in energy should be based on the standards of sustainable development (P5), where the life product is considered from a social, ecological and economic perspective.

The P5 standard evaluates the goals of the project and its final result, approximate operational duration, methods and means of maintenance of equipment and technological processes [31]. ISO 21500:2012 defines the stages of processes: initiation, planning, implementation, control and monitoring activities, and closure. These stages can occur in different ways and at different time intervals. It is worth monitoring the elements of sustainable development both in relation to the project itself and in relation to its impact on society, the economy and the environment [32, 33].

An indicator of the Sustainable Project Management Standard is Social responsibility, which is mandatory. It stipulates organizational standards regarding employees, the hiring procedure, labor relations, staff health and safety, their level of education and the desire to increase the educational level.

Environmental protection is an indicator of the Sustainable Project Management Standard. In other words, the ecological aspect of sustainable development is placed at a high level, related to the impact of activities, programs and projects on flora and fauna of the natural environment, i.e., on the general ecosystem, land, air and water. It is essential to pay attention to CO_2 emissions, the energy used, the impact of the project on water, its quality and volume of consumption, waste, disposal and reuse.

An indicator of the Sustainable Project Management Standard is finance. The economic profit within the project should strive for maximization; the investment payback period should be minimal, and the rate of return where the current value of cash flows is equal to zero, i.e., all flows received during the project are reinvested at the additional rate of return. It is the most profitable rate at which the project breaks even. An indicator of the Sustainable Project Management Standard is business flexibility, which allows businesses to quickly refocus and change the program, sources of funding, and potential investors.

An indicator of the Sustainable Project Management Standard is the stimulation of the economy. It positively impacts the economy by providing the market segment

An indicator of the Sustainable Project Management Standard is project closure. This Standard P5 that provides information helping in closing the project, obtaining information for future similar projects, key advantages and disadvantages.

According to ISO 21500, the opportunities offered by the organizational strategy are defined as economic justification in the process of project acceptance, project documentation management, product creation process, and support process, which will turn into a result transferred to implementation. In the project implementation process, it is crucial to prepare a report on sustainable development, which contains information on the operation efficiency in terms of economic indicators, environmental protection and social responsibility [37–39]. It is worth noting that many organizations in Ukraine offer a full range of services for implementing solar power plants: consulting services, design and pre-project development services, construction, installation, commissioning and commissioning, and service and warranty services [40–42]. The market offers many services that will provide the best possible solutions for implementing household and commercial solutions [43, 44]. It is worth highlighting the following factors that will objectively affect the development of renewable energy resources, including solar energy [45–47]: Ukraine's further accession to the European green course; provision of permanent state support for the production of green electrical energy; Ukrainian society's demand to reduce CO_2 emissions; Ukrainian society's need for eco-friendly energy. A significant problem of renewable resources in the energy sector is the need for improvement of the development of the wholesale electricity market, the high cost of electricity obtained from renewable energy sources, and the high risks of the end user regarding the increase in rates.

5 Conclusions

Expanding the production of renewable energy resources will require the establishment of green clusters, a green economy, greening of transport, and the production of ecological and green products. The green economy is based on the development of renewable energy resources, increasing energy efficiency, promoting the development of the electric vehicle market, and producing and selling eco-friendly goods and services.

The development of renewable energy sources requires compliance with environmental conditions, compliance with sustainable management standards, financial support, investment justification, legislative incentives and tax support. In Ukraine, commercial banks, private investors, private households and international partners actively cooperate in implementing projects to introduce renewable energy sources and produce solar energy. Numerous scientific studies are dedicated to environmental safety, energy efficiency and the financial feasibility of constructing and launching solar plants.

Thus, according to the analysis of capacity and energy output by supply sources in Ukraine for 2018–2020, we can see a positive trend in producing renewable energy

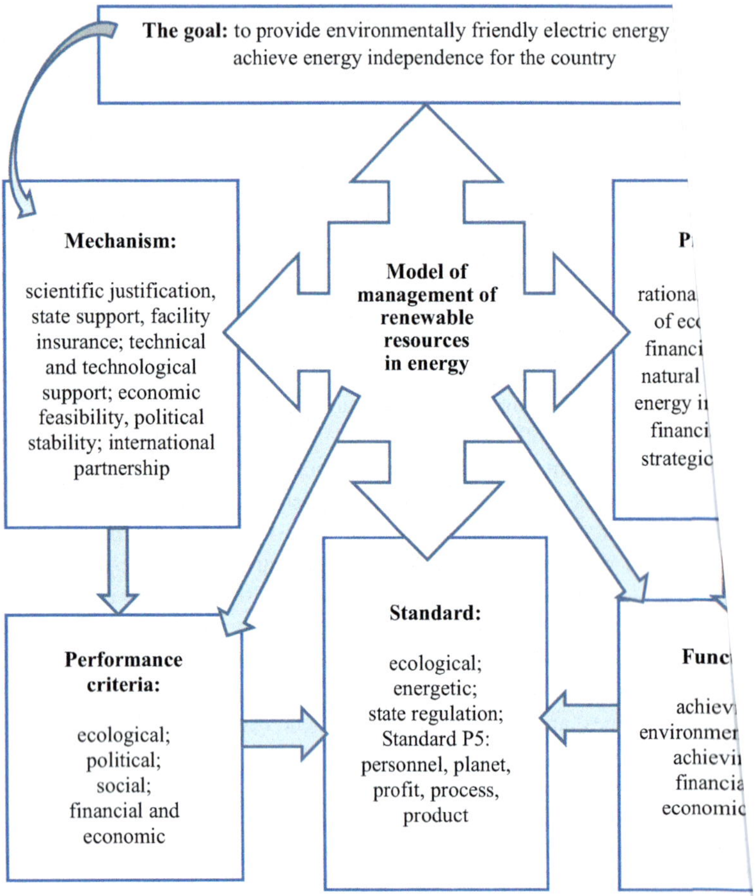

Fig. 5 Model of management of renewable resources in energy

with the necessary products, providing income to the budget, and creating
Thus, the management of renewable resources in the energy sector should on sustainable development standards. The assessment of sustainable deve according to the P5 standard should be carried out during the initial ana forms the main aspects of the project. If risks arise during the introduct implementation of the project, technologies change, and technological pr arise, it is worth performing an analysis and assessment of sustainable devel [34–36]. An indicator of the Sustainable Project Management Standard is management, i.e., restrictions are often used to achieve the desired result. management includes the definition of strategy, policy, measures and goals to the project meet the needs for which it is implemented.

resources. In particular, the installed electric capacity of wind power plants in 2019 compared to 2018 increased from 476 thousand kW to 795 thousand kW and to 1110 kW in 2020. It resulted in a noticeable increase in the volume of electricity supply: from 1760 million kWh in 2019 to 3271 million kWh in 2020. This positive trend is also observed in the operation of solar power plants. Thus, in 2019, the electric power of solar power plants was set at 1953 thousand kilowatts, and in 2020—5194 thousand kilowatts. Along with the increased installed electric capacity, there was a noticeable increase in the volume of electric energy output. Thus, in 2019, the electricity output volume was 1883 million kWh, and in 2020—5684 million kWh.

Considering the specific features of financing renewable resources in energy, it is crucial to determine the volume of increased demand for green investments; financing the costs of implementing renewable resources in the energy sector; debt financing; optimize the structure of funding sources; areas and mechanisms of state stimulation of the production of renewable energy resources.

Ukraine has a huge potential for implementing measures for introducing and generating energy from renewable sources. The most promising is solar energy because the climatic conditions and the number and duration of sunny days confirm this. Undoubtedly, the most suitable territories are the south of Ukraine, which unfortunately is currently under occupation, active hostilities are taking place in this territory, and the solar energy infrastructure has been massively destroyed. The developed action plan for restoring the post-war economy and solar energy occupies an important place in Ukraine because the issue of energy independence is acute right now. Considering Ukraine's problems and significant potential, the construction and implementation of solar power plants are at the initial level of development. Therefore, their financial benefits, environmental safety and social significance require further consideration and research.

References

1. State Agency for Energy Efficiency and Energy Saving of Ukraine (2021) Report on the results of stimulation and use of energy produced from renewable sources in Ukraine for 2019–2020. http://www.saee.gov.ua
2. Herneho Y, Liakhova O (2021) Financing the development potential of alternative energy in Ukraine. Efficient Econ 3:36–44
3. Kucheryava I, Sorokina N (2020) Renewable energy in the world and Ukraine as of 2019–early 2020. Hydropower Ukraine 1–2:38–44
4. Shulga M, Ivashchenko A (2017) Foreign experience of stimulating renewable energy (experience of the People's Republic of China and India). Office of Financial and Economic Analysis of the Verkhovna Rada of Ukraine, Kyiv
5. Skorokhod I, Horbach L (2019) Development of the «green» economy in the countries of the European Union. World Econ Int Econ Relat 30(69):17–21
6. Lischuk VI, Lischuk ME, Moskovchuk AT (2017) Use of renewable resources in energy: world strategies and scenarios of energy market development. Econ Forum 2:30–35
7. Riazanova NO (2017) Economic mechanisms of renewable energy development. Econ State 9:58–61

8. Prokopenko O, Chechel A, Sotnyk I, Omelyanenko V, Kurbatova T, Nych T (2021) Improving state support schemes for the sustainable development of renewable energy in Ukraine. Polityka Energetyczna 24(1):85–100. http://doi.org/10.33223/epj/134144
9. Twidell J (2021) Renewable energy resources. Routledge, London
10. Kurbatova T, Sotnyk I, Prokopenko O, Sidortsov R, Tu Y-X (2021) Balancing Ukraine's energy system: challenges under high renewable energy penetration and the COVID-19 pandemic. E3S Web Conf 280:05007. http://doi.org/10.1051/e3sconf/202128005007
11. Lund H (2007) Renewable energy strategies for sustainable development. Energy 32(6):912–919. https://doi.org/10.1016/j.energy.2006.10.017
12. Bielialov T (2022) Risk management of startups of innovative products. J Risk Financ Manag 15(5):202
13. Dudek M, Bashynska I, Filyppova S, Yermak S, Cichoń D (2023) Methodology for assessment of inclusive social responsibility of the energy industry enterprises. J Clean Prod (in press)
14. Malynovska Y, Bashynska I, Cichoń D, Malynovskyy Y, Sala D (2022) Enhancing the activity of employees of the communication department of an energy sector company. Energies 15(13):4701. https://doi.org/10.3390/en15134701
15. Sotnyk I, Hulak D, Yakushev O, Yakusheva O, Prokopenko OV, Yevdokymov A (2020) Development of the US electric car market: macroeconomic determinants and forecasts. Polityka Energetyczna 23(3):147–164. http://doi.org/10.33223/epj/127921
16. Kennet M (2012) The green economics reader. Green Economics Institute, United Kingdom
17. Galushkina TP, Musina LA, Potapenko VG (2017) The main principles of the implementation of the «green» economy model in Ukraine: training. Manual. Institute of Environmental Management and Balanced Nature Management, Kyiv
18. Decree 605-r of the Cabinet Ministry of Ukraine of August 18 2017: Energy strategy of Ukraine until 2035: «security, energy efficiency, competitiveness». http://mpe.kmu.gov.ua/minugol/doc catalog/document?id=245213112
19. Efimova GV (2002) Evaluation of the economic effectiveness of energy saving investments in industry: author's abstract... candidate. Economy Science, special: 08.07.01. UDMTU Odesa 18
20. Bahaj AS (2002) Means of enhancing and promoting the use of solar energy. Renew Energy 27(1):97–105
21. Ehsanul K, Pawan K (2018) Solar energy: potential and future prospects. Renew Sustain Energy Rev 82:894–900
22. Order 902 of the Cabinet Ministry of Ukraine of October 1 2014: about the National action plan for renewable energy. http://www.zakon4.rada.gov.ua/laws/show/902-2014-%D1%80
23. Institute of Renewable Energy (2020) Renewable energy and energy efficiency in the 21st century. https://www.nas.gov.ua/UA/Book/Pages/default.aspx?BookID=0000017004
24. Investments in renewable energy sources are a step into the future. USELF. EBRD (2022). https://www.eea.europa.eu/signals/signals-2022/articles/a-future-based-on-renewable-energy
25. Omelchenko V (2022) Renewable energy sector before, during and after the war. Nov 11. https://razumkov.org.ua/statti/sektor-vidnovlyuvanoyi-energetyky-ukrayiny-do-pid-chas-ta-pislya-viyny
26. State Statistics Service of Ukraine (2021) Energy balance of Ukraine. https://ukrstat.gov.ua/operativ/menu/menu_u/energ.htm
27. Wärtsilä Energy (2020) Wartsila Energy «Path to 100% renewable energy system in Ukraine». Kyiv
28. Bevs S (2007) Energy efficiency and renewable energy sources. NAS of Ukraine, Kyiv. Ukr. encyclopedia knowledge
29. Kaliappan K, Sankar M, Karthikeyan B, Vineeth B, Raju VC (2019) Analysis of solar energy technology in leading countries. Int J Power Electron Drive Syst 10:1995. http://doi.org/10.11591/ijpeds.v10.i4.pp1995-2004
30. Chamola V, Sikdar B (2016) Solar powered cellular base stations: current scenario, issues and proposed solutions. IEEE Commun Mag 54:108–114

31. Commission of the European Communities, Com (2008) 0019. Proposal for a Directive of the European Parliament and of the Council on the promotion of the use of energy from renewable sources. Brussels
32. Kumar CJ, Majid MA (2020) Renewable energy for sustainable development in India: current status, future prospects, challenges, employment, and investment opportunities. Energy, Sustain Soc 10(2):1–36. https://doi.org/10.1186/s13705-019-0232-1
33. Environmental program (UN): United Nations Environment Program (UNEP) (2020) Energy finance. https://www.unep.org/resources/factsheet/united-nations-environment-programme-unep-energy-finance
34. Ghiani E, Pisano G (2018) Impact of renewable energy sources and energy storage technologies on the operation and planning of smart distribution networks. Oper Distrib Energy Resour Smart Distrib Netw 2:25–48. https://doi.org/10.1016/B978-0-12-814891-4.00002-3
35. Furkan D (2011) The analysis on photovoltaic electricity generation status, potential and policies of the leading countries in solar energy. Renew Sustain Energy Rev 15:713–720. Devyatkina S, Shkvarnytska T (2006) Alternative energy sources: education. Manual. National Aviation University, Kyiv
36. Dudolad A, Kolbushkin Y, Kostin Y (2008) Energetics of quality of life: monogr. Hark. National University of Radio Electronics, Kharkiv. INJEK
37. Zholonko M (2008) Practical energy ecology. Part 1. Alternative sources of energy are the sun and wind. CHNU named after B. Khmelnytskyi, Cherkasy
38. Kondili E (2012) Special wind power applications, vol 2. Technological Education Institute of Piraeus, Athens, Greece, pp 725–746. https://doi.org/10.1016/B978-0-08-087872-0.00225-0
39. Malkin E, Pisarev V (2001) Energy situation in Ukraine and energy saving. National University of Architecture and Architecture, Kyiv
40. Nowak S, Aulich H, Bal JL, Dimmler B, Garnier A, Jongerden G (2006) The European photovoltaic technology platform. In: IEEE photovoltaic energy conversion conferences, pp 2485–2489
41. Hutsaliuk O, Bondar I, Sereda N, Babych O, Shchoholieva I (2021) Organization and management of the development of ecological tourism in a circular economy. E3S Web Conf 255:01026. http://doi.org/10.1051/e3sconf/202125501026
42. Koval V, Borodina O, Lomachynska I, Olczak P, Mumladze A, Matuszewska D (2022) Model analysis of eco-innovation for national decarbonisation transition in integrated European energy system. Energies 15(9):3306. https://doi.org/10.3390/en15093306
43. Koval V, Arsawan IWE, Suryantini NPS, Kovbasenko S, Fisunenko N, Aloshyna T (2023) Circular economy and sustainability-oriented innovation: conceptual framework and energy future avenue. Energies 16(1):243. https://doi.org/10.3390/en16010243
44. Ivashchenko A, Kovalenko Y, Hubanov O, Hubanova T, Lutsenko I, Fialkovska A (2021) Digital tools in cross-cultural analysis of sme financial state support in the conditions of pandemic crisis. J Inf Technol Manag 13:142–162. http://doi.org/10.22059/jitm.2021.82614
45. Ostapenko O, Olczak P, Koval V, Hren L, Matuszewska D, Postupna O (2022) Application of geoinformation systems for assessment of effective integration of renewable energy technologies in the energy sector of Ukraine. Appl Sci 12(2):592. https://doi.org/10.3390/app12020592
46. Shmatko N, Bondar I, Tokareva V, Kovalenko N, Klenin O, Kotlubai V (2021) Project management as a development tool for greening of enterprises of the national economy. IOP Conf Ser Earth Environ Sci 915:012037. https://doi.org/10.1088/1755-1315/915/1/012037
47. Mkhitaryan N, Machulin V (2006) Problems of energy development of Ukraine. Renewable and unconventional energy. Sci Innov 2:63–75

Effectiveness of Bioenergy Management and Investment Potential in Agriculture: The Case of Ukraine

Andrii Dankevych, Iryna Perevozova, Vitalii Nitsenko, Liubov Lozinska, and Yuliia Nemish

Abstract The resource and energy potential of crop biomass forms significant reserves for the economy. There are 42.3% oil, 37.4% meal, 18% husk and 0.7% sunflower oil sludge from 1 ton of sunflower. Currently, scientists are paying a lot of attention to biomass processing and improving ecology. Phosphatide concentrate and lycetin are obtained from sunflower oil sludge. Husk is a source of energy, and its energy value is 18 MJ/t. The purpose of this study is to assess the bioenergy and investment potential in agriculture. In particular, on the example of sunflower, the reserves for increasing the efficiency of electricity production. In particular, the energy cost equivalents of sunflower husk replacement of energy costs in production were estimated. The legal, organizational, and economic foundations of activities related to the prevention or reduction of waste generation have been studied. The substantiation of the necessity of including organic waste in the cycle of substances in the system "soil–plant–person" is confirmed. The need for funds for the modernization of the industry was estimated.

Keywords Agricultural management · Renewable energy · Bioenergy · Bioresource · Sunflower husk · Investment

A. Dankevych
National University of Food Technology, 68, Volodymyrska Str., Kyiv 01601, Ukraine

Zhytomyr Polytechnic State University, 103, Chudnivska Str., Zhytomyr 10005, Ukraine

I. Perevozova · V. Nitsenko (✉) · L. Lozinska · Y. Nemish
Ivano-Frankivsk National Technical University of Oil and Gas, 15, Karpatska St., Ivano-Frankivsk 76019, Ukraine
e-mail: vitaliinitsenko@onu.edu.ua

V. Nitsenko
SCIRE Foundation, Aleja Jana Pawła II 27, 00867 Warsaw, Poland

V. Koval and P. Olczak (eds.), *Circular Economy for Renewable Energy*, Green Energy and Technology, https://doi.org/10.1007/978-3-031-30800-0_6

1 Introduction

The government of Ukraine has included bioenergy in the strategic direction of providing the country with an environmentally friendly type of renewable energy. This is explained by the country's high dependence on imported energy resources and the great potential of available bioenergy. Having a developed biofuel production industry, the country is still far from fully utilizing its resource potential compared to EU countries [1, 2].

At the same time, innovations in the field of waste processing and production of finished products contain significant reserves for increasing the efficiency of the economy.

Ultra-processed products and sunflower waste in the form of organic fertilizers can help reduce the anthropogenic impact on nature, including overcoming such negative phenomena as soil acidification caused by excessive doses of mineral fertilizers and reducing the accumulation of radionuclides in plants.

In this connection, plants are looking for an alternative to calcium and potassium, in the absence of which they absorb other available elements, namely strontium and cesium. It should be noted that the Chornobyl disaster caused the contamination of about 48 thousand square kilometers in Ukraine with radionuclides 137Cs and 90Sr [3].

Analysts note that during the next 50 years fossil fuel reserves may run out [4]. In these conditions, alternative energy sources are a perspective for the development of the economy. Ukraine has the greatest technical potential for the introduction of renewable energy sources (RES) in Europe. The total potential of electricity production from RES is more than 1 million GWh per year. The largest share (85%) is taken by wind energy. The capacity potential of renewable energy sources of Ukraine is estimated at 408.2 GW, including biomass—15.1 GW (4%) [5].

The Government of Ukraine adopted a number of draft laws aimed at stimulating the substitution of natural gas with alternative fuels, in particular:

- Granting preferential status to such projects.
- Simplification of the procedure for concession, lease and lifting of the ban on privatization of heat facilities of communal ownership.
- Development of technical conditions for acceptance of biomethane into the gas transmission system of Ukraine, a mechanism for stimulating its production and consumption.
- Amendments to the Laws of Ukraine regarding the transition to alternative fuels and to the stimulating regulation of relevant business entities [6].

The current agro-industrial production in Ukraine is based on the concept, which is grounded on the costly principle of its development [7]. Hence unrestricted, irrational, and unwise use of productive forces and land resources follows. In Ukraine, the cost of production per unit of gross agricultural output is, in particular, 3–6 times higher and the cost of fuel 2–5 times higher than in Europe [8, 9]. The driving factors for the development of renewable energy sources around the world are the cheapening of technologies. By 2035, the cost of installing solar power plants is predicted to decrease by 50–56% [10, 11].

A specific feature of Ukraine that allows launching projects on an industrial scale is the presence of agricultural holdings [12, 13]. Their accumulation of a wide range of resources, from financial to labor, allows investments in bioenergy. Increasing the efficiency of the interaction of different types of RES in electric networks can be achieved by automating the management of RES modes and optimizing their connection schemes, taking into account their basic features [14].

2 Methodology

Agro-production is characterized by reversibility, cyclicality and uncertainty, as well as by a significant influence of natural and climatic factors. The calculations of forecast indicators of production, yield and area of sunflower crops until 2035 are based on a combined forecast systematized on the basis of research by Ukrainian scientists and the Ministry of Economy of Ukraine [15], which were conducted on the basis of multivariate forecasting using geostatistical, abstract-statistical and system-statistical forecasting methods.

The basic Box-Jenkins time series model [16] was used to forecast production volumes. This model is widely used in economics to predict the results of future periods using data sets from past periods to the present. The advantage of the model is that it implements an autoregressive approach using the so-called "moving average", which allows us to describe both stationary and non-stationary sets (linear vectors) of statistics. In the scientific literature, such a model is also called ARMA (p, q) model.

As you know, the data implementation algorithm consists of two blocks. These blocks are called AR-block and MA-block according to the mathematical apparatus they use. The AR part is responsible for the stationary component of the time series, and the MA part is responsible for random harmonics. In analytical form, the model can be presented as:

$$y_t = \sum_{i=1}^{p} a_i y_{t-i} + \sum_{i=1}^{q} b_q \varepsilon_{q-i} \tag{1}$$

where y_t is the predicted parameter for the period t (years), which corresponds to the indexation of a natural number with the index $i = 1, 2, 3 \ldots . p$; ε_{q-i} is a component that depends on $\boldsymbol{q}$ harmonics of influences; a_i, b_q are coefficients of AR and MA-parts.

For the needs of the presented scientific research the task of development of the strategic forecast for a term long enough has arisen. Therefore, to form a relevant forecast, the traditional model needed to be modified. The purpose of the modification was to level the influence of high-amplitude harmonics of random oscillations in these time series of past periods. Random fluctuations of high-amplitude harmonics for long-term forecasts form a certain initial deviation from the relevant forecast, the effect of which increases with each subsequent period, which leads to significant deviations for remote values (large values of indices i).

Therefore, the algorithm is as follows. Using the autoregressive approach and the moving average, a polynomial equation is formed for each of the sets (vectors) of data. Next, there is the analysis of the amplitudes of random harmonics. In the set of amplitudes, irregular high-amplitude components whose values are higher than a certain, expertly established, threshold value are distinguished by the method of comparison. The next step of the algorithm is to re-apply the traditional algorithm to the specified, but already modified data sets (vectors). Algorithm, namely: predict the area of the raw material base (sunflower crops), million ha (Eq. 2), the gross production of products (seeds) of sunflower, million tons (Eq. 3) and the yield of sunflower, t/ha (Eq. 4).

Accordingly [16]:

$$y = -0.117x^3 + 0.812x^2 + 0.626x + 14.03 \tag{2}$$

$$y = 0.153x^3 - 1.26x^2 + 3.62x + 3.5 \tag{3}$$

$$y = -0.052x^3 + 0.42x^2 - 0.851x + 3.1 \tag{4}$$

where y is the value of the named forecast factor, x is the corresponding time period according to its index i = 1, 2, 3… [16]. Equations (2–4), obtained at the first stage of using the algorithm, were modified into linear equations at the next stage, respectively:

1. To determine indicators of the area under sunflower, million hectares:

$$y = 2.191x + 13.16 \tag{5}$$

2. To determine indicators of gross sunflower production, million tons:

$$y = 0.48x + 5.8 \tag{6}$$

3. To determine the yield rate of sunflower, t/ha:

$$y = 0.12x + 2.47 \tag{7}$$

The equations obtained in the process of research were used by us in further analytical calculations of the balance of sunflower production on the territory of Ukraine. The main parameters used were sown area, gross harvest and yield, economic indicators are the price of grain and profitability. During the growing season there are changes in the size of sown areas, which are primarily associated with crop losses due to changes in weather and climatic conditions.

3 Results

Agricultural land of Ukraine is 32.5 million hectares, it is the 8th bank of agricultural land on the planet. Arable land accounts for 54% of the total area of Ukraine, of which more than 80% have no serious restrictions. This ensures a more efficient farming operation, as it requires less capital and operating costs [17]. Sector specialists suggest that average field size in Ukraine is between 150 ha, while average field size in the EU is below 100 ha. Larger fields allow using large-scale agricultural machinery and equipment, which reduces fuel and labor costs per ha adding to efficient farming [18, 19]. Ukraine is the only country in the world with 2/3 of its total area in mollisols while the remaining 1/3 is in alfisols. Mollisols (chernozem, black soil) are among some of the most fertile and productive agricultural soils in the world. South of Russia, Argentina, North of Kazakhstan, part of Canada and the USA are also covered with mollisols. Due to its geographical location and fertile soils, Ukraine is well positioned to benefit from the advantage in the global commodities market (Fig. 1).

More than 30% of the world's production and export of sunflower oil is concentrated in Ukraine (Fig. 2). Ukraine's climate is changing and becoming warmer. The agro-climatic zones of Ukraine (Steppe, Forest-Steppe and Polissya) have shifted 200 km to the north over the last two decades. 19% of the arable land, which is 6.3 million hectares, became suitable for sunflower cultivation, as a result of which, as of 2000, the area of sunflower cultivation increased by 4.7 million hectares, or four times compared to 1990. Warming increases the yield of oilseeds. According to preliminary estimates, by 2050 the yield will increase by 32.4% [7]. In 2021, the sown sunflower area may reach a record 6.9 million hectares, and the sunflower harvest—17.8 million tons [22]. Under favorable conditions, production may reach 22 million tons by 2050.

Sunflower processing produces a significant amount of by-products. From 1 ton of sunflower, it is possible to obtain 420 kg of oil, 370 kg of meal, 180 kg of husk, as well as a small amount of phosphatide concentrate (Fig. 3).

If one interprets the data of Fig. 3 the calorific value of 2.5 million tons of sunflower husk emits as much heat as 1.0 billion m^3 of natural gas [26, 27]. Taking into account the costs of maintaining cogeneration plants—2.5 million tons of husks will give 3.8 billion kWh in 2020, with further growth to 6.0 in 2050. For comparison, this amount covers 40.2% of the annual electricity needs of the capital of Ukraine.

Given the depletion of basic resources, this is a real step towards the import substitution of natural gas, which took about 41% in the structure of primary energy consumption in Ukraine in recent years, while it is only 21% in the world. As a result, oil extraction plants, which produce sunflower husks and oil sludge as industrial residue, have begun to look for the most profitable ways to use and dispose of them. Payback of such projects is up to 3 years [23, 28, 29]. According to the results obtained [24], the energy consumption for the production of 1 ton of sunflower oil can be equated to 215 kg of sunflower husks (Fig. 4). It should be noted that at the moment about half of the received sunflower husk waste is transformed into electricity and heat.

Geographical location

fertile soil: 2/3 of Ukraine are covered with black earth
climate: moderate continental climate is favourable for agriculture
location: proximity to largest net importing markets: soft commodities (the EU, Middle East)
Highly favorable terrain for agriculture - 80% of farm land is without major constraints (i.e. flat)

Developed infrastructure facilities

Ukraine is capable of transshipping circa 26 million tons of grains and oilseeds per annum through its deep water sea and river ports, which makes farming/grain trading more profitable than in many other CEE countries
well-developed railway network provides cost efficient transportation from storage facilities to ports and to some customers in the EU
Ukraine has installed storage capacity of over 36 million tons, which is steadily growing

Ukraine offers cost leadership in farming

large scale operations enable use of highly efficient farming machinery, thus reducing CAPEX and OPEX per ha
fertile soil reduces consumption of fertilisers
major costs components are lower compared to the EU (fuel, labour, farmland rent, services)

Hidden efficiencies

crop yields are well behind the EU ones
given sufficient investment in proper farming technologies and working capital needs, the farming operations in Ukraine can provide outstanding returns on invested capital on a comparable basis
Ukraine this year plans to achieve record high harvest since 1990 and quintuple its exports of grains and oilseeds

Fig. 1 Ukraine's farming sector: investment case [20, 21]

Compliance with environmental emission standards during combustion implies the presence of powerful electro-filtration. Manufacturers are not interested in complying with environmental emission standards and investing in the environment as these investments are usually unaffordable for them.

The cost of filtration, as a rule, exceeds the cost of the boilers themselves. Payback of such projects is not less than 5 years, and profitability is 12–15%. As a result, the husk, which is pelleted and burned, is sold to third parties. It can be seen that the consumption of heating pellets will only grow, taking into account the increase in the cost of natural gas [30, 31].

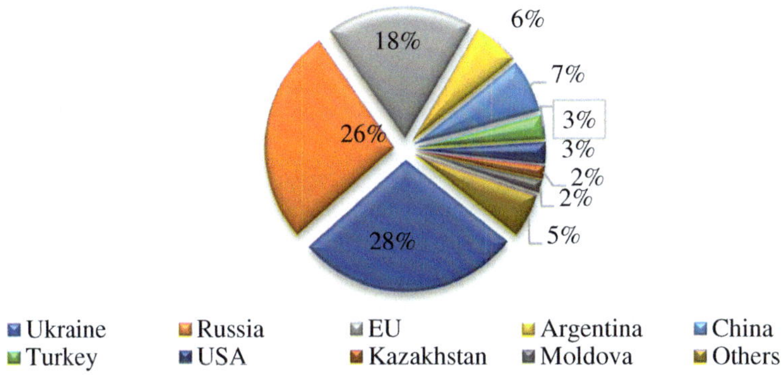

Fig. 2 The place of Ukraine in the world cultivation of sunflowers, 2020 [23, 24]

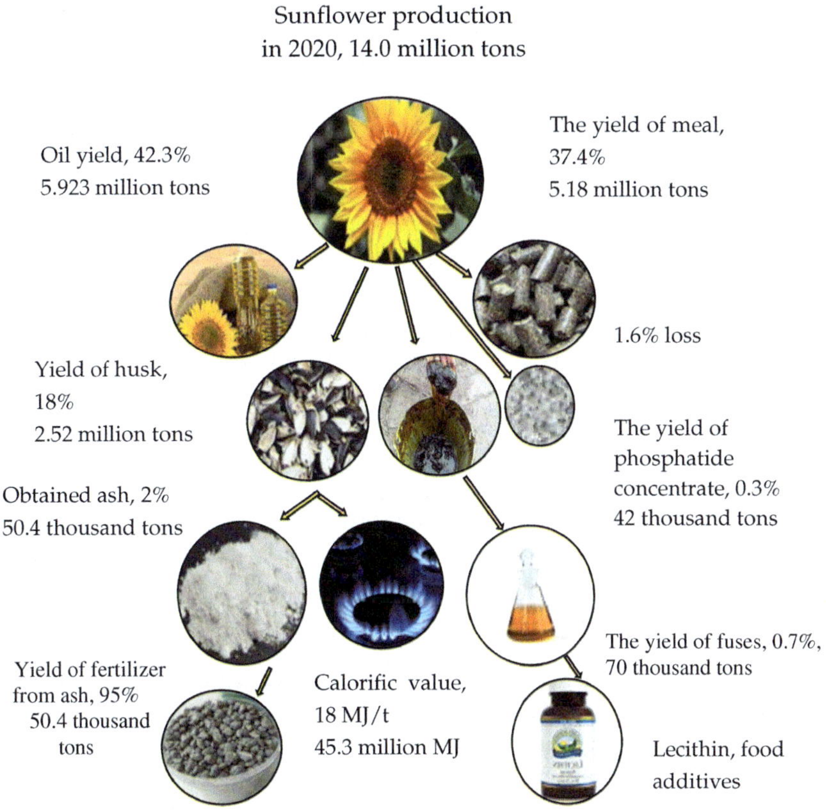

Fig. 3 Bioresource potential of sunflowers in 2020 [24, 25]

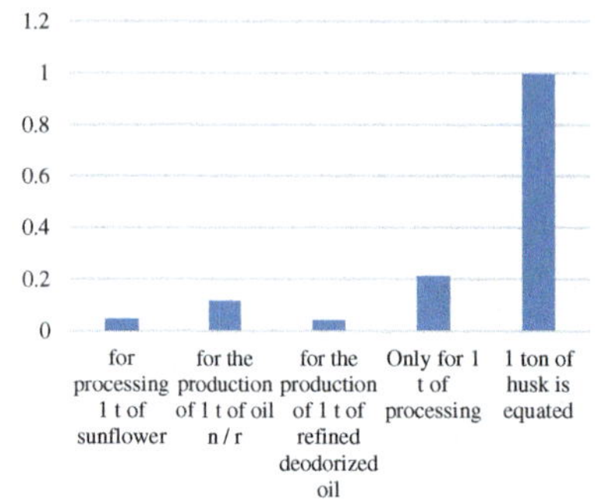

a) The cost of husks for processing products, t.

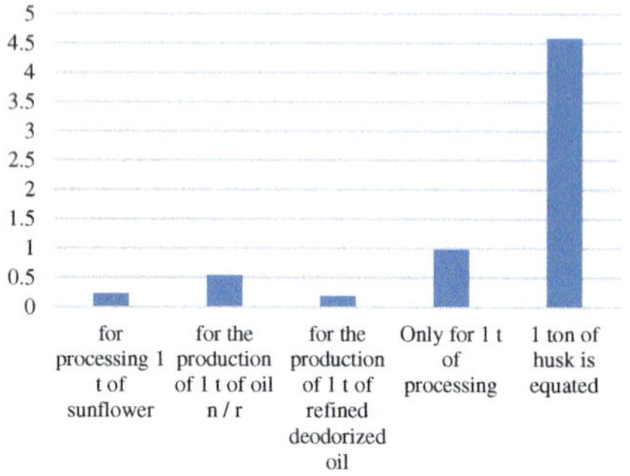

b) Volumes of steam is equivalent to sunflower husk, t.

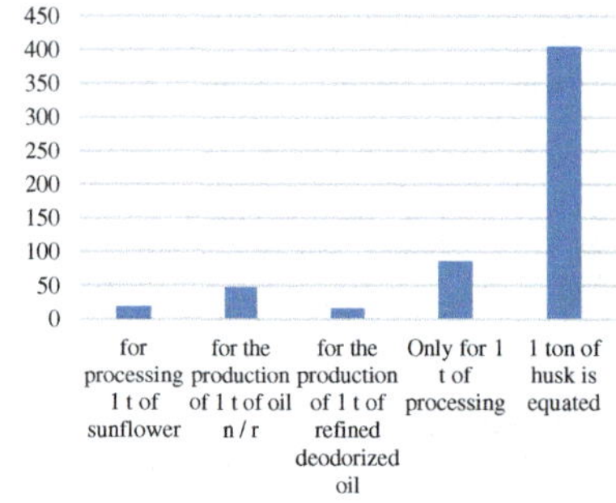

c) Volumes of gas is equivalent to sunflower husk, m^3.

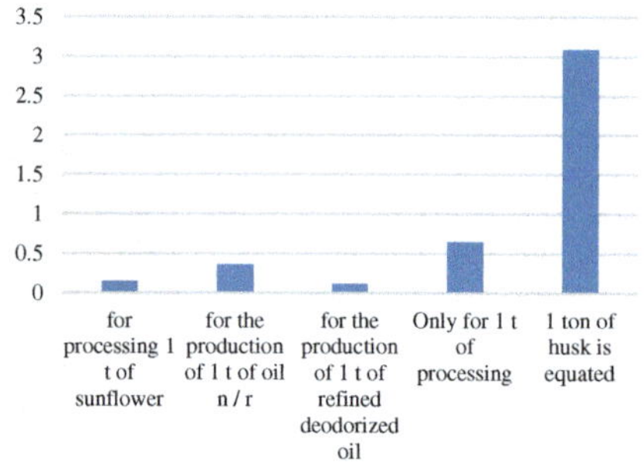

d) Volumes of thermal energy is equivalent to sunflower husk, Gcal.

Fig. 4 The cost of husks for processing products and energy consumption for the production, 2020 [24, 25]

Ensuring the growth of production of renewable energy sources, including electricity, is supported in Ukraine by the "green" tariff mechanism. In particular, since 2009 in Ukraine, its size has been set at the level of the retail tariff for consumers of the second voltage class. With a gradual decrease of 30% until 2028 (Fig. 5) [32, 33].

Among the most important projects in this area, the activities of the company "Kernel" should be noted. Almost all oil extracting plants will be equipped with combined heat and power plants that will generate electricity from burning sunflower husks. Annually, the company's enterprises will produce more than 338 thousand MWh of electricity by burning 228 thousand tons of sunflower husk [33].

Kernel Agroholding already gets more than 50% of energy from renewable sources in the 2021, and its contribution to EBITDA is $10 million in the first half of 2021 [33]. According to the company's semi-annual financial report, the total investment in the program reached $169 million since its inception in 2018, three cogeneration facilities have already been built and their commissioning is being completed, and

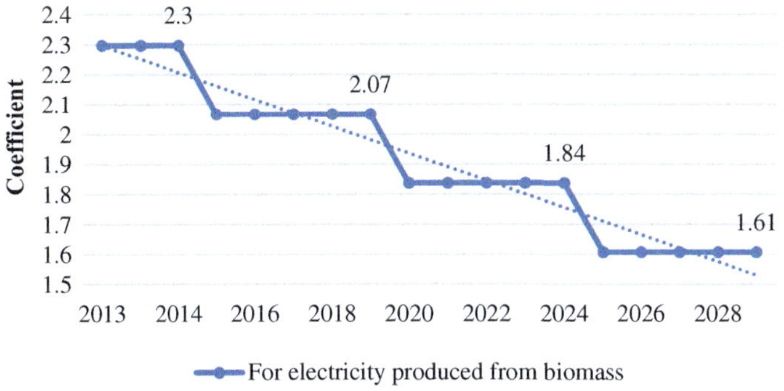

Fig. 5 Coefficient of "green" tariff for facilities put into operation [33]

three more facilities will be commissioned in 2022 [34]. In total, eight projects related to the transition to alternative fuels have been implemented in the country (Table 1).

Biomass plays a significant role in the energy balances of industrialized countries. The use of electricity produced from biofuels at an average annual rate of over 8% was noted in China, Japan, Germany and the United Kingdom. Ukraine has the most energy-intensive economy. The energy intensity of Ukraine's national income is 4–6 times higher than that of the United States, Japan and Western Europe. Consumption of conventional fuel per capita in our country is about 6.5 tons unlike only 4.2–5.5 tons in these countries. The main types of energy resources in Ukraine are coal, oil, gas, nuclear and hydropower (Fig. 6).

The largest energy consumers are the People's Republic of China (30%), the USA (25%), the Russian Federation (about 10%), Japan (5.5%) and Germany (7.3%). The energy consumption of Ukraine (about 3%) was comparable to the energy consumption of Great Britain, France, Canada or India [38]. As of the end of 2020, renewable energy capacities have been installed, which produce about 7.3% of the total electricity supplied. The largest share of RES in Ukraine is occupied by wind and solar power plants, which in 2016 produced 925 GWh and 492 GWh of electricity, respectively (Table 2).

Biomass is the most progressively growing segment of renewable energy sources in Ukraine (Table 3). The development of the use of biomass energy is important not only in terms of electricity production, but also even more valuable from the standpoint of ecology, as it solves the problem of waste disposal of industrial and agricultural enterprises.

If in 1995 the share of biomass in the total consumption of primary energy in the European Union was 3%, in the US it was 3.2%, in Denmark it was 8%, in Finland it was 21%, in Austria it was 11%, and in Sweden it was 19% [40], then the IEA forecasts an increase in demand for primary energy from biomass to 12% of the total global demand for primary energy by 2030. The energy potential of biomass in Ukraine is estimated at 11–18 million tons of conventional fuel per year, which is equivalent to about 14.3–24% of total primary energy consumption. In fact, in 2020, this share will be 0.05% in Ukraine. Until recently, its use was limited to

Table 1 Chronology of implementation of projects on transition to alternative fuel of oil enterprises of Ukraine

Name of the enterprise	Year	Project implementation stage
OJSC "Melitopol MEZ"	1997	Development of design documentation for reconstruction, installation of steam boiler KE 10-14 MT
CJSC "Pologi SEZ"	1998	Design and commissioning works on gas generator automation. Reconstruction of E 35-3,9-440 GM for husk burning
OJSC "Vinnytsia MZhK"	2000	Commissioning and thermochemical tests of the modernized steam boiler KE-10-1,4-285 on husk and natural gas
OJSC "Kirovogradolia"	2000	Development of technical documentation for the reconstruction of the steam boiler DKVR 20-13-250 Art. №3 when transferring to the burning of sunflower husk
CJSC "Poltava MEZ - Kernel Group"	2000	Development of design documentation for reconstruction and installation of steam boiler E-16-2,1-350 HMDV for burning sunflower husk with auxiliary equipment
CJSC with II "Dnipropetrovsk SEZ"	2000	Commissioning works, ecological and thermal tests, adjustment of a water mode on a steam boiler of production of LARDET-BABCOCK firm with fuel "sunflower husk" with a productivity of 30 t/h
CJSC "Zaporizhzhya MZhK"	2001	Commissioning and research work of the steam boiler OL-20, which burns sunflower husk (manufactured by RAFAKO)
OJSC "Chernivtsi MZhK"	2002	Reconstruction of a boiler room with installation of a reconstructed boiler with a steam capacity of 12 t/h for burning sunflower husk

Source Formed by the authors using [35]

direct combustion over an open fire or in furnaces and stokers with relatively low efficiency and did not comply with environmental requirements for emissions of harmful substances [39].

According to the Biomass Research Center, of the 109 million tons of agricultural waste (straw, corn stalks, sunflower waste, cattle and pig manure, poultry manure, food waste, etc.), 59 million is used as fertilizer or for livestock needs and only 1 million tons is used for the production of heat and electricity. The analytical report of the International Finance Corporation estimates that using only 20 million tons of waste for energy production, Ukraine will cover 25% of annual electricity needs, which will replace more than 8.02 billion m^3 of natural gas. That is, the use of only agricultural residues makes it possible to refuse from the import of natural gas by 75–80%. The analysis of experts showed that it is agricultural pellets from sunflower

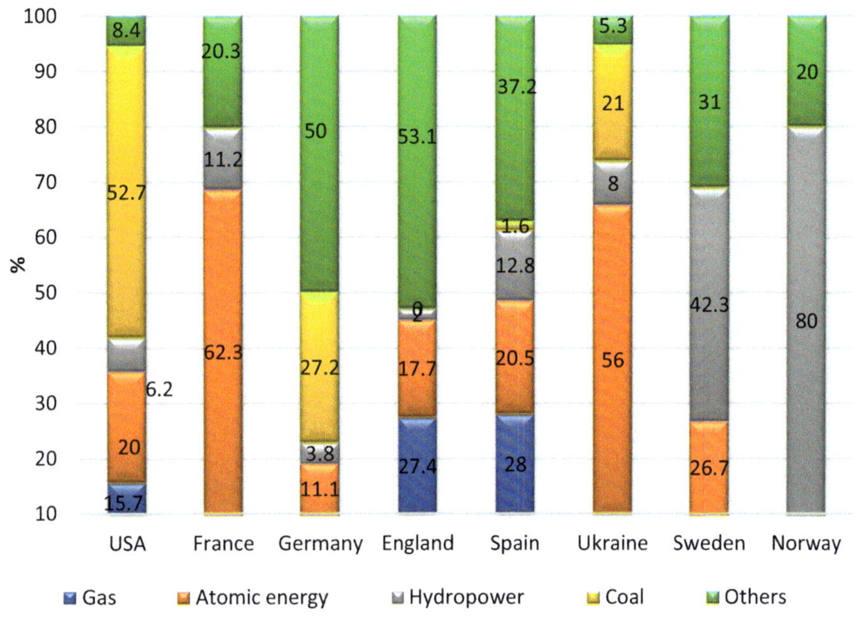

Fig. 6 Structure of electricity production in the world, 2019 (in %) [35–37]

Table 2 Electricity generation by type of power plant

Type	2010 (%)	2012 (%)	2014 (%)	2019 (%)	2020 (%)
NPP	47.4 ▼	45.5 ▼	48.5 ▲	53.9 ▲	51.2 ▼
TPP	41.5 ▲	44.7 ▲	41.3 ▼	36.2 ▼	35.2 ▼
HPP/PSP	6.9 ▲	5.5 ▼	5.0 ▼	5.1 ▼	5.1 ▲
Block stations	4.2 ▲	4.0 ▼	4.3 ▼	1.09 ▲	1.21 ▲
SPP/WPP/biomass	–	0.3 ▲	1.0 ▲	3.62 ▲	7.3 ▲
Incl. biomass			0.055 ▲	0.019 ▲	0.023 ▲

Source Calculated by the authors using [37, 39]

husks that have a greater impact on the growth of production volumes in the Ukrainian biofuel sector. Expert estimates show that the energy potential of biomass in Ukraine may increase to about 45 million tons of o.e./year (Fig. 7) by 2050. It is expected that the total installed capacity of bioenergy equipment will increase to 49.6 GW and 5.2 GW by 2050. The total consumption of biofuels will be more than 20 million tons of o.e./year.

The implementation of these measures will require investments of 21… 33 billion euros and will lead to the replacement of about 20.1 billion m^3 of natural gas a year and the creation of more than 160.2 thousand jobs by 2050 [38].

Table 3 Renewable energy capacity, MW

Type	2010	2012	2014	2019	2020
Wind	87	194	426	1170	1314
Solar	3	326	411	4925	6094
SPP of households	–	–	0.1	553	779
Small hydropower	68	73	80	114	109
Biomass	–	6	35	55.9	91
Biogas	–	–	14	70.3	103
Large HPPs/PSPs	5400.2	5400.2	5724.2	6048.2	
Total	5558.2	5999.2	5853.3	12,936.4	8490

Source Calculated by the authors using [37, 39]

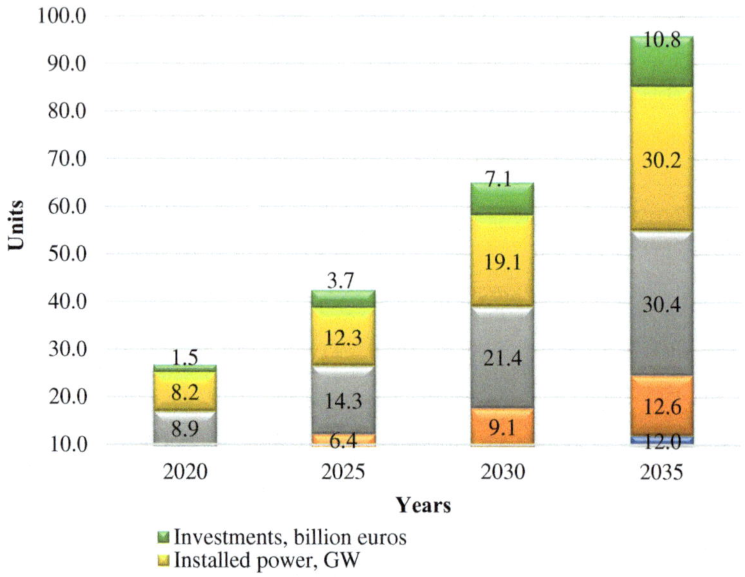

Fig. 7 Forecast of bioenergy development until 2035 [36, 37]

The pioneers in this direction are agricultural holdings, which have invested significant funds in the modernization of existing facilities and construction of new bioenergy ones. According to preliminary estimates, there are more than 90 large agricultural holdings in Ukraine, which control about 27.8% of arable land and are the largest producers of grain and industrial crops.

Agricultural holdings usually cover the entire cycle of production, processing and sale of products and provide targeted reproduction of the industry on a new technological, organizational and economic basis [40]. The raw material base of Ukraine allows the production of about 40,114 tons of phosphatide concentrate per year, at the same time, at the moment, this potential is being developed by only 28.1%, about 10,512 tons of the finished product suitable for sale are sold (see Fig. 3). Projected investments in the modernization of the industry are estimated at 35–55 million dollars [2]. In terms of the amount of carbon dioxide emissions when burning sunflower husks, it can be equated to natural gas, while (ash content) does not exceed 3.1% [41]. It is possible to solve the disposal of ash formed after burning by producing complex fertilizers from them.

Granulated ash is an environmentally friendly fertilizer suitable for use in organic production. When using 1 ton of granular ash, 90 kg of calcium, 50 kg of magnesium, 40 kg of sulfur, 260 kg of total potassium can be added to the soil [2]. The current potential of ash from sunflower husks in Ukraine is estimated at 50,283 tons. Approximate investments in the modernization of the industry are estimated at 30.4 million dollars.

4 Discussion

The agro-industrial production of Ukraine needs changes in the policy of environmentally unlimited, irrational and unbalanced use of production resources, the absence of an ecological culture of production. Of particular concern is the extremely unsatisfactory ecological condition of agricultural lands. Over the past three decades, the area of eroded arable land in Ukraine has almost doubled. At the same time, researchers [42–45] note that the functioning of the system "soil–plant–animals" necessitates the inclusion of waste in the cycle of substances. Researches confirm the need to ensure ecologically clean and rigid parameters of the obtained electricity when burning sunflower husks [46, 47]. To implement these tasks, scientists recommend using an automated control system for all technological production cycles.

The restraining factors for entering the market are: limited information on the assessment of the regional technical potential of various types of RESs; outdated Soviet energy systems and limited possibilities of connection to them; complex permission system. Also guarantees of buyout of energy produced and guarantees of imposing a "green" tariff at the planning stage of the project are not provided. A characteristic feature of the grain production process in Ukraine is a sharp increase in yield variance in recent years. According to some authors, this trend is due to the influence of meteorological factors, which undergo similar changes. This behavior of the system increases the uncertainty and risk of investment decisions. If the probability of obtaining critically low yields becomes significant, the investor may decide to change the investment plans. For this reason it is necessary to have an assessment in the annual perspective. The most successful predictions are realized when an adequate mathematical model of the object is made. Methods aimed at making the model are divided into two large groups: (1) development of linear stochastic models. This area has received a special name "system identification". This approach has acquired the most complete form in the works of Box and Jenkins, who proposed the ARMA model [48, 49]. Significant capital investments at the first stages restrain the development of RES bioenergy [50]. According to some researchers, it is possible to increase the efficiency of renewable energy sources in electrical distribution networks by choosing the optimal power of asynchronous generators and compensating facilities, choosing the optimal connection schemes for RES, in automation of management and optimization of its work as a part of electric system that allows to reduce losses of the electric power in electric networks, and also to reduce prime cost and to increase volumes of the generated electric power [51–53]. The roadmap for the development of bioenergy in Ukraine until 2050 should take into account [54, 55]:

- growth of the share of bioresources in the structure of solid biofuel consumption up to 60% and 20% of the total volume, respectively, in 2050;
- minimal growth in the use of wood biofuels 1.2 times in 2050;
- a significant increase in the production of biogas and liquid biofuels up to 4.7 million tons of oil equivalent per year in 2050;
- launch and growth of production of biomethane and motor biofuels of the second generation up to 2.4 million tons of oil equivalent per year and 0.43 million tons of oil equivalent per year, respectively, in 2050.

5 Conclusions

Studies show that Ukraine uses no more than 10% of its bioenergy potential of sunflower husks. At the same time, the existing bioresource potential of only sunflower husks can provide light for about 5 million people. Investments in obtaining alternative electricity from sunflower husks are estimated at 1.9 billion dollars.

The creation of agricultural holdings is a model for the further development of the agricultural sector of Ukraine's economy, which will bring it to a qualitatively new trajectory of stable, highly efficient and competitive operation that will allow solving priority tasks. In addition, this strategy is designed to revive domestic agricultural production and to ensure an ecologically balanced transformation of rural areas and settlements. Studies have proven the feasibility of using sunflower husk as an alternative fuel for the production of heat energy and electricity on this basis for the production of granular organic fertilizers from ash. Implementation of these measures on a comprehensive basis, on an industrial scale, will solve environmental, economic and energy problems of the industry: recycle waste, obtain alternative energy sources, increase profitability, ensure organic production of environmentally friendly fertilizers and increase the manufacture of organic products.

To realize competitive advantages, a constructive, more predictable state policy in the sphere of RES is needed, which should become a priority in the system of economic transformations in the conditions of the global economic crisis. First of all, it concerns the strategy design for the development of the sector; improvement of legislation and settlement of property relations; free access to credit for the development of market infrastructure and the elaboration of a range of motives for investors, especially those who invest in innovative projects for its development.

References

1. Bioenergy is a strategic direction of Ukraine (2021). https://www.em-eco.net.ua/biblioteka/stati/szhiganie-luzgi-podsolnechnika
2. Havrysh V, Nitsenko V, Hruban V (2022) Sorghum-based power generation in Southern Ukraine: energy and environmental assessment. Agriculture 12:2148. https://doi.org/10.3390/agriculture12122148
3. Balanovska T, Mykhailichenko M, Holik V, Dramaretska K, Troian A (2021) Development management of agricultural enterprises. Financial and credit activity: problems of theory and practice, vol 3, no 38, pp 134–143. http://doi.org/10.18371/fcaptp.v3i38.237434
4. Geletukha G (2017) Current state and prospects of bioenergy development in Ukraine. Bioenergy Association of Ukraine, Kyiv
5. Vyhovska H (2004) The structure of waste in Ukraine, their sources and quantitative indicators. Native Nat 3–4:23–25
6. Andreichenko A, Andreichenko S, Smentyna N (2021) Ensuring biosphere balance in the context of agricultural waste management. Philos Cosmol 26:46–61. http://doi.org/10.29202/phil-cosm/26/4
7. Rozin V (2021) From engineering and technological process to post-cultural technology. Future Human Image 15:99–109. http://doi.org/10.29202/fhi/15/9
8. Dankevych A (2017) Analysis of the investment potential of enterprises in the agricultural sector of the economy. In: Problems of theory and methodology of accounting, control and analysis, vol 3, pp 37–42. http://doi.org/10.26642/pbo-2017-3(38)-37-42
9. Dehodiuk EH, Mamontov VT, Hamalei EH (1988) Ecological bases of fertilizer use. Urozhay, Kyiv
10. Biliavskyi M (2020) Guidelines for the development of alternative energy in Ukraine until 2030. https://razumkov.org.ua/statti/oriientyry-rozvytku-alternatyvnoi-energetyky-ukrainy-do-2030r
11. Kucher A, Hrechko A (2021) Assessment and analysis of regional features of resource-saving land use for sustainable management. Sci Pap Ser Manag Econ Eng Agric Rural Dev 21(1):431–441
12. Kuznetsova A (2010) Straw use in Ukraine—opportunities and options. Policy paper series [AgPP No. 31]. Kyiv
13. Rahman M, Chowdhury S, Mohammad Zayed N, Ali Imran M, Hanzhurenko I, Nitsenko V (2022) Does globalization trigger an ecological footprint? Rocznik Ochrona Środowiska 24:141–162. http://doi.org/10.54740/ros.2022.011
14. Dankevych A, Sosnovska O, Dobrianska N, Nikolenko L, Mazur Y (2021) Ecological and economic management of innovation activity of enterprises. Naukovyi Visnyk Natsionalnoho Hirnychoho Universytetu 5:112–118. http://doi.org/10.33271/nvngu/2021-5/118
15. Erickson P, Lazarus M (2018) Would constraining US fossil fuel production affect global CO_2 emissions? A case study of US leasing policy. Clim Change 150:29–42. https://doi.org/10.1007/s10584-018-2152-z
16. Box GEP, Jenkins GM, Reinsel GC, Ljung GM (2015) Time series analysis: forecasting and control, 5th edn. Wiley, Hoboken, New Jersey
17. Nitsenko VS, Havrysh VI (2016) Enhancing the stability of a vertically integrated agro-industrial companies in the conditions of uncertainty. Actual Probl Econ 10(184):167–172
18. Horák J, Bilan Y, Dankevych A, Nitsenko V, Kucher A, Streimikiene D (2023) Bioenergy production from sunflower husk in Ukraine: potential and necessary investments. J Bus Econ Manag 24(1):1–19. https://doi.org/10.3846/jbem.2023.17756
19. Chowdhury S, Rahman MM, Raisa TS, Akhter S, Zayed NM (2020) A nexus among air pollution, energy consumption and agricultural productivity in Bangladesh: an ARDL model with structural break. Int J Manag 11(7):93–103
20. Mishchenko V, Vyhovska H (2006) Waste management: standards as a management tool. Stand Certification Qual 4:37–40

21. Geletukha GG, Zheliezna TA, Drozdova OI (2013) Comprehensive analysis of technologies for energy production from solid biomass in Ukraine. Part 1. Industrial heat engineering, vol 3, pp 56–63
22. Nitsenko V, Mardani A, Streimikis J, Shkrabak I, Klopov I, Novomlynets O, Podolska O (2018) Criteria for evaluation of efficiency of energy transformation based on renewable energy sources. Montenegrin J Econ 14(4):253–263. http://doi.org/10.14254/1800-5845/2018.14-4.17
23. Salkova IY, Polishchuk NV, Broyaka AA, Kucher AV (2020) Cost effectiveness and attracting investments in crop sector. Financial and credit activity: problems of theory and practice, vol 1, no 32, pp 176–186. http://doi.org/10.18371/fcaptp.v1i32.200355
24. Lystopad VL, Kukhta VH (2010) Sunflower husk market in Ukraine, and implementation of projects for the production of solid fuel in the fat and oil industry. Oil Fat Complex 2(29):13–17
25. Dubrovin VO (2004) Biofuels (technologies, machines and equipment). TsTI «Enerhiia ta elektryfikatsiia», Kyiv
26. Korolchuk Y (2013) Renewable energy: prospects for Ukraine. Radio Svoboda. https://www.radiosvoboda.org/a/25044801.html
27. Comparative table of calorific value of some types of fuel (2021). http://a-invest.com.ua/aktualno/tablitsa-teplotvornosti
28. Perea-Moreno M-A, Manzano-Agugliaro F, Perea-Moreno A-J (2018) Sustainable energy based on sunflower seed husk boiler for residential buildings. Sustainability 10:3407. https://doi.org/10.3390/su10103407
29. Bala-Litwiniak A, Zajemska M (2020) Computational and experimental study of pine and sunflower husk pellet combustion and co-combustion with oats in domestic boiler. Renew Energy 162:151–159. https://doi.org/10.1016/j.renene.2020.07.139
30. Bala-Litwiniak A (2021) Environmental and economic aspects of combustion of biomass pellets containing a waste glycerol. Combust Sci Technol 193(11):1998–2008. https://doi.org/10.1080/00102202.2020.1746774
31. Zaporozhets A, Babak V, Sverdlova A, Shcherbak L, Kuts Y (2022) Review of the state of air pollution by energy objects in Ukraine. Syst Res Energy 2(71):42–52. http://doi.org/10.15407/srenergy2022.02.042
32. Kuznetsova A, Kutsenko K (2010) Biogas and "green tariffs" in Ukraine—is it profitable to invest. AgPP No. 26
33. Lysohor M (2021) In Poltava, KERNEL built a 10 MW biopower plant running on sunflower husk. Poltavshchyna. https://poltava.to/news/59568/
34. Smirnov OA, Ladyhina OA (2008) Research of the state and trends of energy development in Ukraine and ways to overcome its problems with the use of renewable energy sources. Machinery in agricultural production, industrial engineering, automation, vol 20, pp 315–320
35. Geletukha GG, Zheliezna TA, Drahniev SV, Bashtovyi AI (2020) Potential and prospects for using agribiomass for energy in Ukraine. Thermophys Therm Power Eng 42(1):42–51. http://doi.org/10.31472/ttpe.1.2020.5
36. EWEA executive summary analysis of wind energy in the EU-25. European Wind Energy Association
37. European Commission (2015) Report from the Commission to the European Parliament, the Council, the European Economic and Social Committee and the Committee of the Regions. Renewable energy progress report 117
38. Koval V, Sribna Y, Mykolenko O, Vdovenko N (2019) Environmental concept of energy security solutions of local communities based on energy logistics. In: 19th international multidisciplinary scientific GeoConference SGEM, vol 19, no 5.3, pp 283–290. http://doi.org/10.5593/sgem2019/5.3/S21.036
39. Yaroshchuk O (2018) For the green reserve—annual results of alternative energy in the agro-industrial complex and plans for 2018. https://agropolit.com/spetsproekty/396-zeleniy-rezerv--richni-pidsumki-alternativnoyi-energetiki-v-apk-ta-plani-na-2018
40. Danyliuk V, Riepina I, Shafalyuk O, Kovylina M, Nitsenko V (2020) Functional and investment strategies of technical development of enterprises. Naukovyi Visnyk Natsionalnoho Hirnychoho Universytetu 3:115–121. http://doi.org/10.33271/nvngu/2020-3/115

41. Masnyak O et al (ed) (2019) Glossary of technical terms in the field of energy efficiency and renewable energy sources. Lviv Polytechnic, Lviv
42. Ostapchuk T, Orlova K, Biriuchenko S, Dankevych A, Marchuk G (2021) Defuzzification in the process of managerial estimating the value of agricultural lands. Agric Resour Econ Int Sci E-J 7(4):62–81. http://doi.org/10.51599/are.2021.07.04.04
43. Kougioumtzis M-A, Kanaveli I-P, Karampinis E, Grammelis P, Kakaras E (2019) Combustion of olive tree pruning pellets versus sunflower husk pellets at industrial boiler. In: XXIV international symposium on combustion processes, Sept 23–25, Wroclaw, Poland
44. Papp V, Beszédes S, Szalay D (2019) Mechanical and energy examination of different agripellets. Analecta Technica Szegedinensia 13(1):40–47. http://doi.org/10.14232/analecta.2019.1.40-47
45. Safin RR, Ilalova GF, Khasanshin RR (2020) Study of pyrolysis of annual crop refuse under reduced pressure. Solid State Phenom 299:974–979. https://doi.org/10.4028/www.scientific.net/ssp.299.974
46. Le Clef E, Kemper T (2015) Sunflower seed preparation and oil extraction. In: Martínez-Force E, Dunford NT, Salas JJ (eds) Sunflower chemistry, production, processing, and utilization. Academic Press and AOCS Press. http://doi.org/10.1016/C2015-0-00069-7
47. Spirchez C, Lunguleasa A, Croitoru C (2019) Ecological briquettes from sunflower seed husk. E3S Web Conf 80:01001. http://doi.org/10.1051/e3sconf/20198001001
48. Ion IV, Popescu F (2017) Improving the energy balance in a sunflower oil mill. Scientific works of university of food technologies, vol 64, no 1, pp 164–171
49. Chowdhury S, Hasan KBMR, Rahman MM, Islam KMA, Zayed NM (2020) Causal relationship among carbon dioxide (CO_2) emissions, renewable energy, population and economic growth in Bangladesh: an empirical study. Res World Econ 11(6):196–206. https://doi.org/10.5430/rwe.v11n6p196
50. Dankevych A (2019) The efficiency of the use of production resources in the conditions of corporatization. AgroInCom 7–9:122–125
51. Murru M, Calvo CL (2020) Sunflower protein enrichment. Methods and potential applications. OCL 27:17. https://doi.org/10.1051/ocl/2020007
52. Koval V, Sribna Y, Kaczmarzewski S, Shapovalova A, Stupnytskyi V (2021) Regulatory policy of renewable energy sources in the European national economies. Energy Policy J 24(3):61–78. http://doi.org/10.33223/epj/141990
53. Ostapenko O, Olczak P, Koval V, Hren L, Matuszewska D, Postupna O (2022) Application of geoinformation systems for assessment of effective integration of renewable energy technologies in the energy sector of Ukraine. Appl Sci 12(2):592. https://doi.org/10.3390/app12020592
54. Geletukha GG (2021) What will be the bioenergy of Ukraine in 2050. Ukrainian Pravda. https://www.epravda.com.ua/columns/2021/04/30/673528/
55. Gluba Ł, Rafalska-Przysucha A, Szewczak K, Łukowski M, Szlązak R, Vitková J, Kobyłecki R, Bis Z, Wichliński M, Zarzycki R, Kacprzak A, Usowicz B (2021) Effect of fine size-fractionated sunflower husk biochar on water retention properties of arable sandy soil. Materials 14(6):1335. https://doi.org/10.3390/ma14061335

Towards Overcoming Energy Crisis and Energy Transition Acceleration: Evaluation of Economic and Environmental Perspectives of Renewable Energy Development

Olga Ostapenko, Gulzhan Alina, Madina Serikova, Lyudmila Popp, Tetiana Kurbatova, and Zamira Bashu

Abstract In the study, assessing the prospects for using renewable energy sources is considered a means of overcoming the energy crisis and increasing environmental security in the concept of the clean energy transition. The study determined the values of final energy consumption for various energy resources and sources of thermal and electrical energy in Ukraine for 2015–2019. The structure of final energy consumption by industry in Ukraine in 2019 was evaluated. The values of global investments in traditional and renewable energy for the period 2019–2021 are estimated and forecast data of total primary energy demand from various sources for the European Union, according to the "EU Basic Scenario", for the period up to 2060, as well as forecast values of total primary energy demand. The approach proposed in the article makes it possible to reasonably determine the prospects for the use of environmentally safe energy-saving technologies with the use of non-traditional and renewable energy sources in Ukraine in the foreseeable future until 2060, based on the estimated trends in the reform of the energy sector in the world and the European Union, which should be considered as a means of overcoming the energy crisis and increasing environmental security in the clean energy transition.

O. Ostapenko (✉)
Vinnytsia National Technical University, 95 Khmelnytskyi Highway, Vinnytsia 21021, Ukraine
e-mail: ostapenko@vntu.edu.ua

G. Alina
Kazakh University of Economics, Finance and International Trade, 7 Zhubanov St., 020000 Nur-Sultan, Kazakhstan

M. Serikova · Z. Bashu
L.N. Gumilyov Eurasian National University, 2 Satpaeva St., 010000 Nur-Sultan, Kazakhstan

L. Popp
Toraighyrov University, 64 Lomova St., 140008 Pavlodar, Kazakhstan

T. Kurbatova
Sumy State University, 2 R.-Korsakova St., Sumy 40007, Ukraine

University of Sussex, Falmer, Brighton BN1 9RH, UK

V. Koval and P. Olczak (eds.), *Circular Economy for Renewable Energy*, Green Energy and Technology, https://doi.org/10.1007/978-3-031-30800-0_7

Keywords Renewable energy sources · Energy sector · Energy security · Clean energy transition

1 Introduction

In 2015, the UN defined 17 global goals for sustainable development, which must be achieved by 2030 [1]. Among these goals, significant attention is paid to combating climate change and strengthening environmental protection. The European Union, in turn, established the "Green Pact", a plan to introduce a circular economy. These steps will help Europe become more ecological and climate neutral by 2050 [1].

The circular economy is being introduced to achieve Sustainable Development Goals (SDGs) 8, 9 and 12. SDG 8 will ensure sustainable, inclusive and sustainable growth, mainly through increasing the global efficiency of resource use in consumption and production. SDG 9, which deals with sustainable infrastructure and industrialization, calls on industries to implement resource-saving and environmentally friendly technologies. SDG 12 refers to sustainable consumption and production based on resource efficiency and waste reduction.

Currently, many countries use a linear type of economy, a predominantly unidirectional model of production and consumption, where natural resources are needed for mass production, and products are often disposed of after one-time use. It becomes evident that the vast majority of modern production and consumption processes could be more sustainable and ecological, resulting in climate changes and an increase in the price of raw materials and the final product. It leads to the exhaustion of resources. It leads to overproduction and overconsumption, which threatens to achieve sustainability.

According to the formulations of the UN, "sustainability" and "environmentalism" are precisely what meet the needs of today without jeopardizing the ability of future generations to meet their needs [2]. In addition, the use of "green" technologies has ceased to be for business exclusively an image characteristic for and directly affects the sustainability and innovative attractiveness of companies.

The circular economy manifests itself in the long term and covers all industrial sectors at the regional, global and local levels. It is also the only option that will reduce consumption and simultaneously extend the service life of products and resources. The circular economy is based on the renewal of resources and is an alternative to the linear economy (creation, use, waste disposal). It should change the classic linear production model, focusing on products and services that minimize waste and other types of pollution [3].

Europe wants to become climate neutral by 2050, outlined in the "Green Deal" of the European Commission [4]. With the help of the «Green Deal», the European Commission demonstrates its willingness to invest in sustainable models of the future and to stop or slow down the pace of global warming. Moreover, the goals of the "Green Deal" not only cover the environmental need for a greener future but also include economic and social considerations that allow Europe to pursue a resource-efficient and future-oriented growth strategy [4].

2 Literature Review

The European Union, which adopted a circular economy course in 2015, expects from the implementation of relevant initiatives a significant reduction in the amount of waste (in particular, greenhouse gas emissions) and an increase in the productivity of resource use. Among the advantages of the closed-loop model in the EU is the reduction of harmful effects on the environment. However, it is about more than just the environmental friendliness of this approach. It also has clear business economic advantages. First and foremost, it is about reducing dependence on imported resources, the prices of which may jump due to fluctuations in the markets, changes in the exchange rate, and these resources may even be unavailable.

Specific areas of action to achieve the targets and, therefore, European climate neutrality can be found in the Green Deal [4]. One of the primary measures mentioned is promoting the more efficient use of resources by transitioning to a clean, closed-cycle economy.

The circular economy has been a much-discussed topic in recent years, and although it has gained widespread acceptance, it has yet to be widely applied. In Germany, for example, the Circular Economy Act (Kreislaufwirtschaftsgesetz, KrWG) has been in force since 2012, which adheres to the principle of waste management that is safe for people and the environment, taking into account technical, economic and social aspects [5].

The circular economy model is an attractive concept for business because it offers many concrete strategies for implementation at the corporate level and allows organizations to focus on resource-efficient production and ensure resource reuse throughout the value chain. The closed cycle economy model increases resource use efficiency and reduces the impact on the environment and natural capital by developing products more suitable for secondary processing, introducing effective technologies and turning waste into a resource.

The closed cycle economy's main principles are restoring resources, processing secondary raw materials, and transitioning from fossil fuels to renewable energy sources. Also, this type of economy is considered part of the Fourth Industrial Revolution [6], due to which the rationality of using resources, including natural ones, will generally increase.

The EU has adopted an action plan for the transition to a cyclical economy, which provides that such a model becomes the basis of the EU's sustainable development strategy and provides for the development of appropriate state regulation. The transition to a cyclical economy has three indisputable advantages: a reduction in the negative environmental impact due to the reduction of the use of resources during production; a reduction of production costs due to a decrease in the number of primary resources used; the emergence of new markets, the creation of new jobs and an increase in the level of well-being.

The closed cycle (or circular) economy changes the traditional linear concept of the economy. In the linear model, products are produced, used and disposed of (take-make-dispose), while the circular approach is based on the 3-R principle:

- Reduce: reduce the use of resources and give priority to renewable materials;
- Reuse: products are used as efficiently as possible;
- Recycle: recover by-products and waste for further use.

Technology is undoubtedly the key to realizing the "principle of three R". At the same time, talking not only about technologies that ensure the safety of disposal of household and industrial waste or increase the possibilities of recycling secondary resources but also ensure the improvement of the efficiency of technologies along the entire chain: extraction of raw materials–transportation–production–consumption.

The ultimate goal of the policy based on the "three R principle" is a model of society where all resources (minerals, energy, and water) will be used so efficiently that the very concept of "waste" will cease to exist. The economy begins to work within the framework of the Zero waste concept [7, 8].

There are three key features of the circular economy:

1. Enhanced control over natural resource stocks and maintaining a stable balance of renewable resources to preserve and maintain an inexhaustible level of natural capital.
2. Optimization of consumption processes through developing and distributing products, components and materials that correspond to the highest level of their reuse.
3. Identify and prevent negative external effects of current production activities from increasing the efficiency of economic and environmental systems.

As part of the circular economy, the following options for business models are possible, which can be used both separately and in combination:

- circular chains of added value—a model in which fully renewable sources replace limited resources;
- increasing the life cycle of the product—a model that allows preserving the economic benefit for as long as possible with the help of restoration, repair, modernization or re-marketing of the product;
- exchange and joint consumption (sharing economy)—a model based on the exchange of goods or assets that have a low utilization rate;
- product as a service (servicing)—a model where customers use the product by "renting" with payment according to the fact of use;
- recovery and recycling—a model that uses technological innovations and opportunities to recover and reuse resources.

Waste processing and reuse are essential elements of the circular economy. The analysis of modern challenges and the development of technologies allows us to add two more to these principles: Redesign and Rethink, reconstruct and rethink:

- Redesign. This principle includes the need to use innovative technologies when creating new products, which, even at the development stage, allow for the possibility of recycling, reusing and minimizing the impact on the environment;
- The principle of Rethink. This principle aims to form ecologically oriented social consciousness, worldview and education. More and more attention is paid to implementing this principle in many countries.

According to data [9], the growth in renewable energy capacity in the world broke the previous record in 2021, and the demand for biofuels in the world has almost recovered to the level before COVID-19. How quickly renewables can replace fossil fuels will depend on many factors.

Examining the latest market and policy changes as of April 2022 [9], it is predicted that 2022 and 2023 will see new global growth in renewable energy capacity and increased demand for biofuels. After the Russian invasion of Ukraine, it is difficult to assess the potential impact of the recently announced targets on the 2023 forecast in the absence of rapid policy implementation. While market uncertainty is compounding the challenges, there is a new emphasis on energy security—especially in the European Union—which is also driving an unprecedented political push to accelerate the adoption of energy efficiency and renewable energy technologies. The forecast of renewable energy markets for 2023 and the following years will depend on whether new and more vigorous policies will be introduced shortly [9].

The momentum from the "net zero pledge" and sustainable financing has yet to translate into a significant increase in spending on clean energy projects worldwide. As highlighted in the IEA's new Roadmap to Net Zero by 2050 [10], the policy should facilitate a significant surge in clean energy investment this decade. Governments, companies and financial institutions have committed to zero emissions by 2050 or soon after. In many countries with developed economies, the financial community has united around sustainable financing, establishing funds and implementing relevant initiatives. Investments in clean energy are growing moderately but remain significantly less than the level needed to avoid severe consequences of climate change. It is estimated that investment in clean energy would need to double in the 2020s to hold temperatures well below 2 °C and more than triple to keep temperature increases stabilized at 1.5 °C.

The world's major economies are promoting innovation and increasing technology funding in pursuit of net zero emissions. Supporting innovation in the energy sector is a critical component of net-zero plans. However, in 2020, state spending on research and development in the energy field continued to grow while the share of low-carbon technologies in the total volume increased to 80%. Signals for investment in low-carbon energy technology innovation as of early 2021 were generally positive. By 2030, more than US $50 billion in public funds could become available for large-scale demonstration projects for large-scale low-carbon energy technologies [10].

Global trends in rapidly rising energy and materials prices and shortages of crucial minerals, semiconductors and other components create potential obstacles to the transition to clean energy [11–14]. The Covid-19 pandemic and Russia's invasion of Ukraine have led to significant disruptions in global energy and technology supply chains. Against this background, the IEA study "Energy Technology Prospects 2023" (ETP-2023) [15] contains an analysis of possible risks and an assessment of opportunities associated with the development and expansion of clean energy supply chains and technologies in the foreseeable future that are considered them through the prism of achieving energy security and sustainability [16–18].

The study aims to assess the prospects for using renewable energy sources in the world, Europe and Ukraine in the foreseeable future until 2060, which is considered a means of overcoming the global energy crisis and increasing environmental security in the concepts of "green" energy transition.

3 Research Methods

In recent years, many scientific studies by various authors have been devoted to the practical application and assessment of the potential of renewable energy sources [19–23]. This research is conducted based on methodological and statistical data from the databases of the International Energy Agency (IEA), which uses the databases of the International Renewable Energy Agency (IRENA), the World Bank, the United Nations Statistics Division (UNSD) and others ("Tracking SDG7", "Rise", "Esmap").

In particular, the following databases of the International Energy Agency (IEA) were used: "World Energy Balance Highlights" (IEA database of energy balances for over 185 countries and regions) [24]; "Energy efficiency Indicators Highlights" (database of energy efficiency indicators) [25]; "World Energy Outlook 2022" (data from analysis of energy development in the context of energy and environmental security) [26]; "World Energy Investment 2021" [contains global and regional supply and end-use investment data; regional data on investment in fossil fuels, renewables, power grids and end-use (energy efficiency, renewables)] [27]; "Global Energy Review: CO_2 emissions in 2021" (contains global indicators of CO_2 emissions and is based on a detailed analysis by the IEA by region and by type of fuel based on official, national and publicly available energy, economic and weather data) [28]; "Greenhouse gas emissions from Energy Highlights" (IEA's annual data on green-house gas emissions from the energy sector) [29]; "SDG7 Database" (contains data on access to electricity and clean cooking (SDG 7.1) and progress on the SDGs on renewable energy (SDG 7.2) and energy efficiency (SDG 7.3) (last updated March 2022) This dataset is updated by the IEA to track progress towards Sustainable Development Goal 7) [30]; "Net zero scenario by 2050" (contains data together with projections at the global level for a scenario of net zero emissions by 2050 and includes projections at the global level for the Net Zero Emissions by 2050 (NZE) scenario, based on detailed modeling of the energy sector) [31].

4 Results and Discussion

The transition of the energy sector to clean energy is causing significant structural changes in electricity generation profiles worldwide. The share of renewable generation has increased dramatically over the last decade. This trend is expected to continue and even accelerate in line with the achievement of climate change goals. It should be noted that traditional power plants, mainly those that use coal, atomic and hydropower, stagnate or decline over time [32].

Renewable energy sources, including solar power, wind power, hydropower, biofuels and others, are at the heart of the transition to lower carbon and more sustainable energy systems. In recent years, the use of renewable energy sources has been overgrown, thanks to political support and a sharp reduction in costs for solar photovoltaic and wind energy in particular [33]. It is predicted that the expansion of renewable energy capacities in the next five years will occur much faster than was predicted just a year ago. It is planned that in 2022–2027 generation from renewable energy sources will increase by almost 2400 GW. According to the central forecast [34], renewable energy sources will account for more than 90% of the global expansion of electric power capacities during the forecast period. China mainly drives the upward revision to the previous forecast; the European Union, the United States and India broadly implemented existing policies and regulatory and market reforms and introduced new regulatory mechanisms faster than expected in response to the energy crisis [34].

While overall growth in renewable energy capacity could help limit global warming to 1.5 °C [35]. Concerns about global energy security caused by Russia's invasion of Ukraine have prompted countries to increasingly turn to renewable energy sources, mainly solar and wind, to reduce countries' dependence on imported fossil fuels, the prices of which have skyrocketed. As mentioned above, the global capacity of renewable electricity is expected to grow by 2400 GW during 2022–2027, which is equal to the total capacity of China today, according to data [35], which is the latest edition of the annual report of the IEA. It confirms that the significant expected growth in global renewable electricity capacity will be 30% higher than the values predicted just a year ago, underscoring how quickly governments have secured additional policies to expand the use of renewable energy sources. The report [35] states that over the next five years, renewable energy sources will account for more than 90% of global electricity growth, overtaking coal and becoming the largest source of global electricity by the beginning of 2025.

It is predicted that the amount of renewable energy added in Europe in 2022–27 will be twice as large as in the previous five years due to a combination of energy security issues and climate ambitions. Even faster deployment of wind and solar photovoltaic systems sectors could be achieved if the EU member states quickly introduced several strategies for optimizing and improving schemes for stimulating renewable energy sources [29].

Faster growth in the use of renewable energy in the foreseeable future will bring the world closer to achieving "net zero emissions" by 2050, providing a chance to limit global warming to 1.5 °C [33]. The IEA and Ukraine are deepening bilateral cooperation with the help of a new joint work program; this cooperation is focused on restoring the energy system of Ukraine and its transition to sustainable energy. The joint work program focuses on Ukraine's critical short- and long-term energy priorities, emphasizing energy security, clean energy transition, and Ukraine's recovery plan. The work areas outlined in the program include energy system security, hydrogen depletion, renewable energy sources, biogas, and cooperation on data and statistics. The two-year program is based on the long history of cooperation between the IEA and Ukraine, which began in 2007. The Agency closely cooperated with Ukraine through the EU4Energy program of the European Commission. It should be noted that Ukraine officially joined the IEA as an association country in July 2022 at the signing ceremony in Warsaw, opening a new page in their long-term partnership [36].

It is predicted that in 2023 the European Union will face a potential shortage of almost 30 billion cubic meters of natural gas. However, the risk of a shortage can be avoided by intensifying efforts to improve energy efficiency, deploying renewable energy sources, installing heat pumps, promoting energy preservation and increasing gas supplies.

The IEA report [37] outlines critical actions to address the potential supply–demand gap if Russian pipeline supplies fall to zero, including faster energy efficiency implementation and renewable energy sources. The IEA's study of government energy spending shows an increase in clean energy funding in response to the global energy crisis, alongside efforts to reduce and afford electricity costs. Global public spending to support clean energy has increased by more than $500 billion since March 2022, as the global energy crisis spurs new policies to reduce dependence on fossil fuels, according to an IEA study [37]. It was determined that countries with developed economies account for almost 95% of the support for investments in clean energy allocated worldwide since the pandemic's beginning. Developing countries have channeled their more limited resources into short-term measures to keep transport, electricity and fuels affordable.

The energy world is developing in the era of clean energy technologies, which creates new risks, forcing countries to develop industrial strategies. According to forecast estimates [37], the world market of key mass-produced environmentally friendly energy technologies will amount to approximately 650 billion US dollars annually by 2030. The study determined the indicators of estimated cumulative energy savings from efficiency for the IEA countries by production sectors for 2015–2019. It was determined that the highest values of estimated cumulative energy savings from efficiency are achieved in the industry [38]. Figure 1 shows the values of final energy consumption by various energy resources and thermal and electrical energy sources in Ukraine for 2015–2019. It shows an increase in the use of non-traditional and renewable energy sources (biofuels and waste), which became possible due to the implementation of green energy policy measures [39] and a decrease in the consumption of traditional energy resources (coal, natural gas), the use of electricity and thermal energy does not change significantly during the studied period.

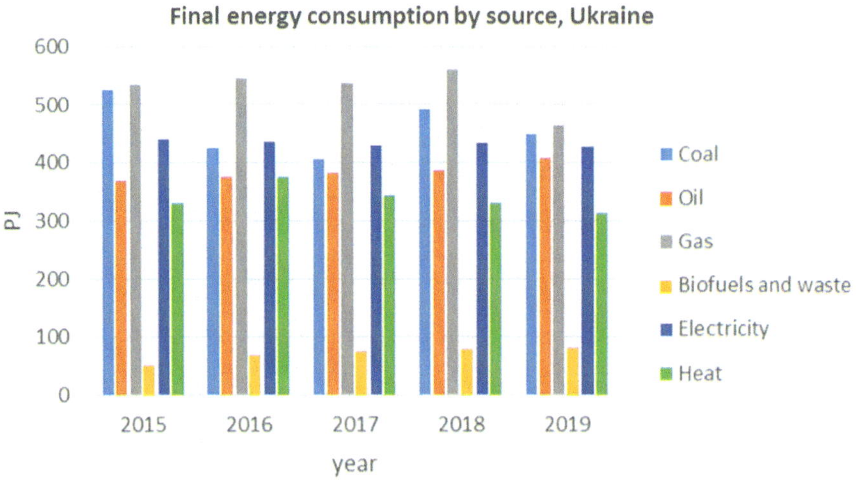

Fig. 1 Values of final energy consumption by various energy resources and thermal and electrical energy sources in Ukraine for 2015–2019 [24–31]

The study estimated the structure of final energy consumption by sector in Ukraine in 2019. It was determined that the most significant final energy consumption traditionally corresponds to the production sector.

Table 1 shows the values of global investments in traditional and renewable energy for 2019–2021. Table 1 shows a gradual decrease in investments in traditional energy and an increase in financing of non-traditional energy.

The energy sector occupies a central place in the fight against climate change. The energy sector accounts for two-thirds of the total volume of greenhouse gas emissions, so the energy sector is a central player in ensuring emissions reduction and climate change mitigation.

In recent times, promoting sustainable development and combating climate change have become intertwined aspects of energy planning, analysis and policy development. Figure 2 shows the percentages of the growth/decrease rate of global CO_2 emissions for the periods of 2019–2020 and 2020–2021 from traditional, non-traditional, and renewable energy sources (biomass and waste) and in technological processes.

Figure 3 shows the percentages of the growth/decrease rate of CO_2 emissions of energy and technological processes in the world, certain countries and the European Union for the periods 2019–2020 and 2020–2021.

Figure 4 shows the percentages of the growth/decrease rate of energy greenhouse gas emissions in the world, certain countries and the European Union for the periods 2019–2020 and 2020–2021.

Figure 5 shows the forecast data of the total primary energy demand from various sources for the European Union, according to the "EU Basic Scenario" for the period until 2060, as well as the forecast (total) values of the total primary energy demand. It shows a gradual decrease in the total primary demand for energy from traditional

Table 1 Values of global investments in traditional and renewable energy for 2019–2021

World	2019	2020	2021
Total (billion $)	1936	1688	1851
Supply (by type)	1643	1399	1531
Fossil fuels (fuel supply and power)	986	726	813
Renewables	345	367	382
Electricity networks	269	259	286
Other supply	43	47	51
End-use	294	289	320
Energy efficiency	268	266	295
Renewables and other end-use	26	23	25
Fuels	864	621	708
Fossil fuels	856	613	694
Oil	478	332	365
Gas	279	191	238
Coal	99	90	91
Low-carbon fuels	9	8	14
Power	778	778	823
Generation	505	513	530
Coal	57	49	45
Gas and oil	74	64	74
Nuclear	39	42	44
Renewables	336	359	367
Battery storage	4	6	7
Electricity networks	269	259	286
Memo: Oil and natural gas upstream	475	326	351

Source Data obtained by the authors based on IEA databases [24–31]

sources (oil, coal, natural gas) and an increase in demand for non-traditional and renewable energy sources. Also, the primary demand for nuclear and hydraulic energy will grow, leading to the insignificant or minimal formation of harmful emissions into the atmosphere. It should be noted that by 2060, according to forecast data, it is planned to reduce the total primary demand for energy from various sources in the European Union to a value of 54,564 PJ.

Figure 6 shows forecast data of fuel energy consumption for the generation of electrical and thermal energy from various sources for the European Union, according to the "EU Basic Scenario", for the period up to 2060. Figure 6 shows that by 2060, the use of oil, coal, and natural gas in electricity and heat generation will gradually decrease, which will be ensured by the gradual increase in the use of nuclear energy, hydropower, as well as non-traditional and renewable energy sources in heat and

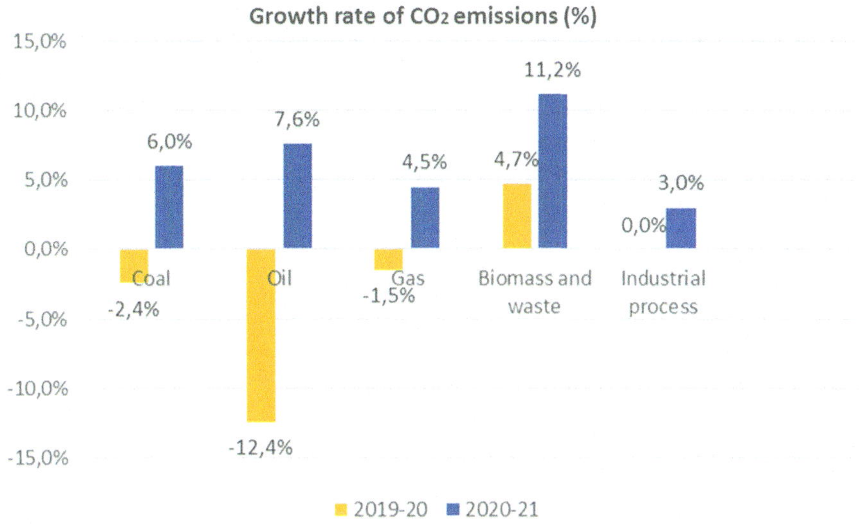

Fig. 2 Percentages of the growth/decrease rate of global CO_2 emissions for the periods 2019–2020 and 2020–2021 from the use of traditional, non-traditional, and renewable energy sources (biomass and waste) and in technological processes [24–31]

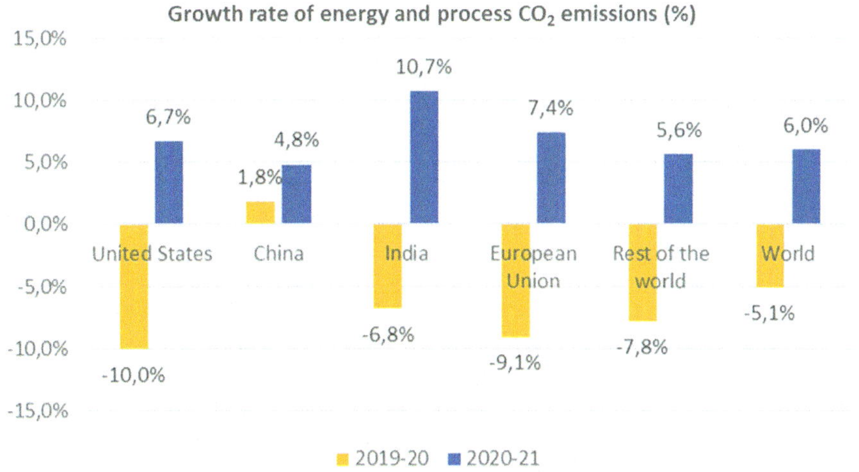

Fig. 3 Percentages of the growth/decrease rate of CO_2 emissions of energy and technological processes in the world, certain countries and the European Union for the periods 2019–2020 and 2020–2021 [24–31]

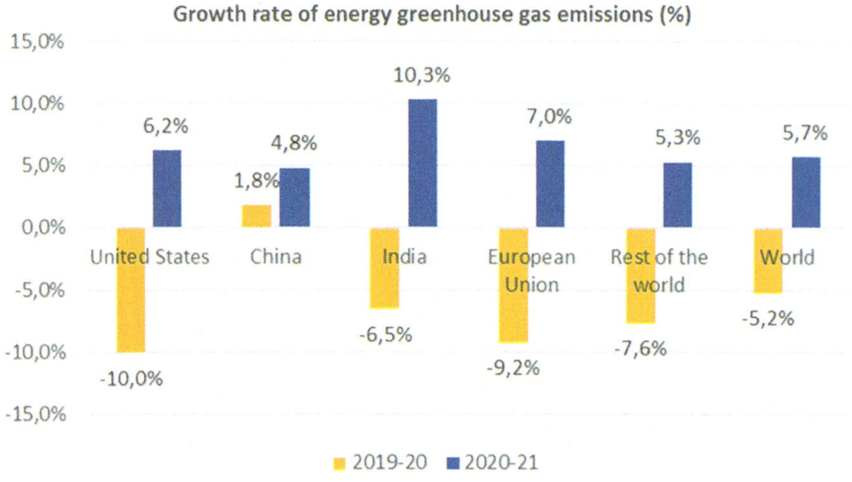

Fig. 4 Percentages of the growth/decrease rate of energy greenhouse gas emissions in the world, certain countries and the European Union for the periods 2019–2020 and 2020–2021 [24–31]

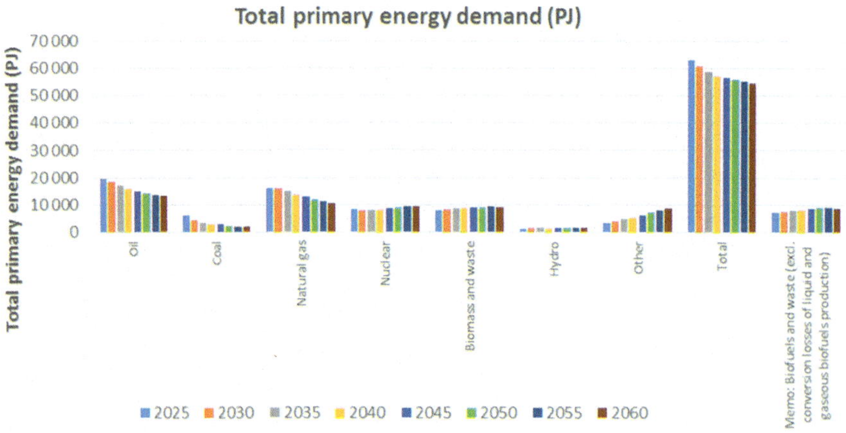

Fig. 5 Forecast data of the total primary energy demand from various sources for the European Union, according to the "EU Basic Scenario", for the period up to 2060, as well as forecast values of the total primary energy demand [24–31]

electricity generation (biomass and waste, geothermal energy, wind and solar energy, ocean energy), the use of hydrogen and other energy sources was not predicted.

Figure 7 shows forecast data of final energy demand from various sources for the European Union, according to the "EU Basic Scenario" for the period up to 2060. It shows that by 2060, the final need for energy from such sources as oil, coal and natural gas will gradually decrease, the final need for thermal energy will decrease and the final need for electrical energy, energy from biomass and waste, energy from

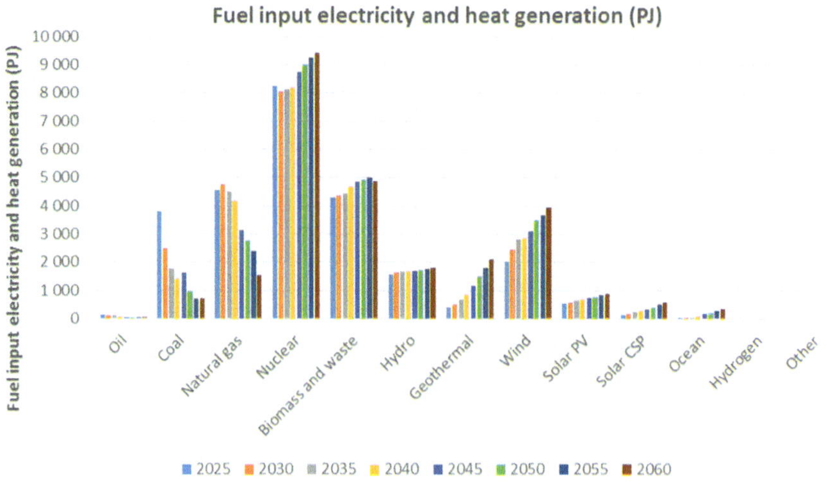

Fig. 6 Forecast data of fuel energy consumption for the generation of electrical and thermal energy from various sources for the European Union, according to the "EU Basic Scenario", for the period up to the 2060 year [24–31]

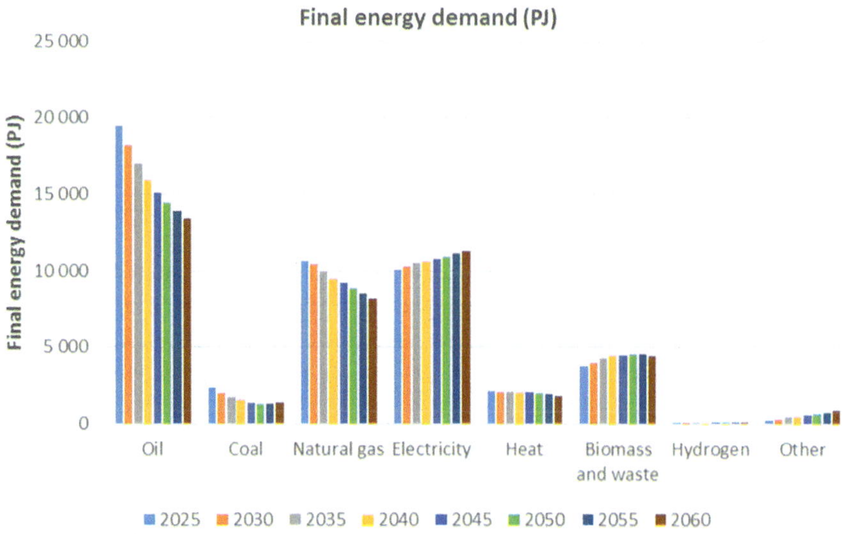

Fig. 7 Forecast data of the final energy demand from various sources for the European Union, according to the "EU Basic Scenario", for the period up to 2060 [24–31]

the day and other new technologies. It should be noted that by 2060, according to forecast data, it is planned to reduce the final need for energy from various sources in the European Union to a value of 41,415 PJ.

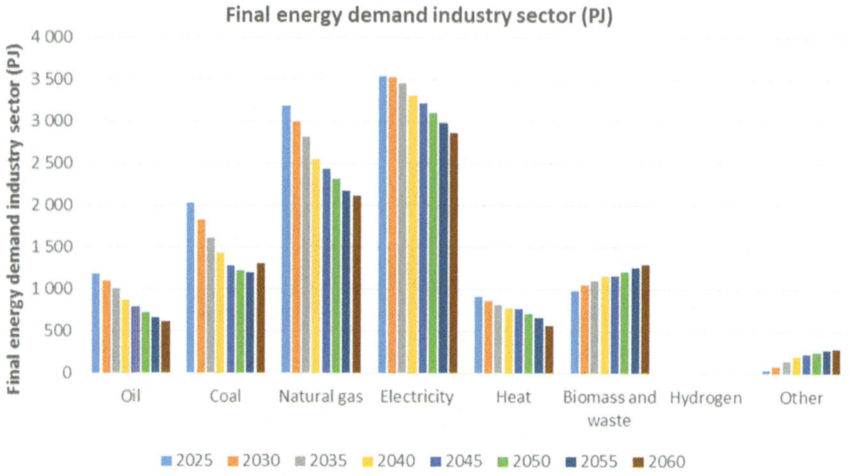

Fig. 8 Forecast data of the final energy demand in the industrial sector from various sources for the European Union, according to the "EU Basic Scenario", for the period up to 2060 [24–31]

Figure 8 shows forecast data of the industrial sector's final energy demand from various European Union sources, according to the "EU Basic Scenario", for the period up to 2060. It shows that by 2060, the final energy demand in the industrial sector from such sources as oil, coal, and natural gas will gradually decrease, and the final demand for thermal and electrical energy in the industrial sector will decrease. However, the final demand for energy from biomass, waste and other new technologies and the use of hydrogen was not predicted. It should be noted that by 2060, according to forecast data, it is planned to reduce the final energy demand in the industrial sector from various sources in the European Union to the value of 9009 PJ.

Figure 9 shows forecast data for the growth of electricity production from various sources for the European Union, according to the "EU Basic Scenario", for the period up to 2060. We observe a gradual decrease in the growth of electricity production from traditional sources and an increase in its generation from non-traditional sources. It shows that until 2060, the amount of electricity production using oil, coal, and natural gas will gradually decrease, which will be ensured by the gradual increase in electricity production using nuclear energy, hydropower, as well as non-traditional and renewable energy sources in electricity generation (biomass and waste, geothermal energy, wind and solar energy, ocean energy), the use of hydrogen and other energy sources was not predicted. It should be noted that by 2060, according to forecast data, a gradual increase in the volume of electricity production in the European Union is planned to the value of 3519 TWh.

Figure 10 shows forecast data for the growth of electric capacity from various sources for the European Union, according to the "EU Basic Scenario", for the period up to 2060. It shows a gradual decrease in the amount of electric capacity from traditional energy sources and an increase in the amount of electric capacity from non-traditional and renewable energy sources. It shows that by 2060, the growth

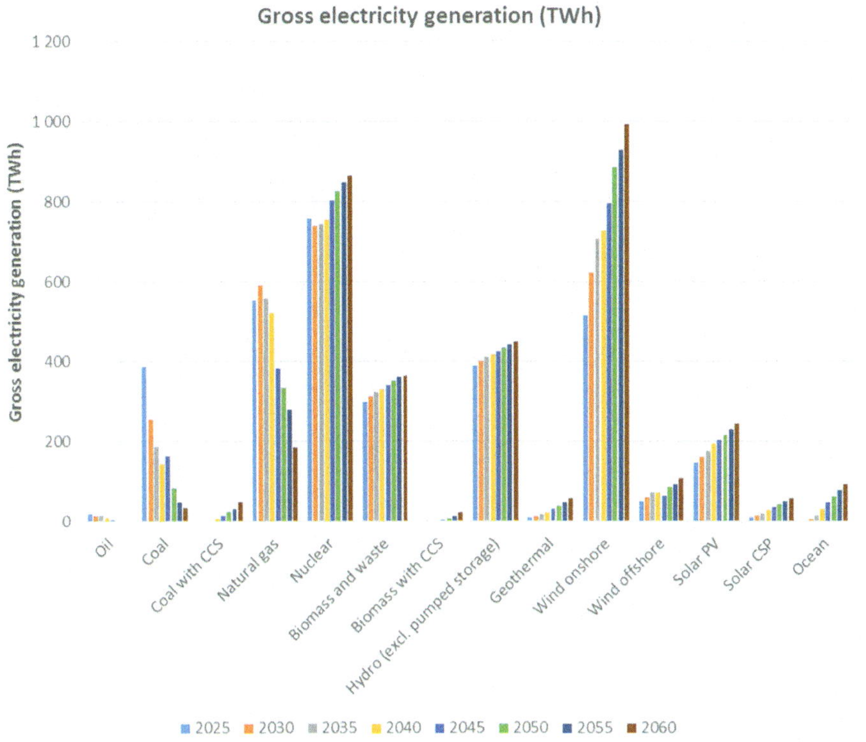

Fig. 9 Forecast data of the growth of electricity production from various sources for the European Union, according to the "EU Basic Scenario", for the period up to 2060 [24–31]

of electric capacity using oil, coal, and natural gas in electric generation will gradually decrease, which will be ensured by the gradual increase of electric capacity using nuclear energy, hydropower, as well as non-traditional and renewable sources of energy in electrical generation (biomass and waste, geothermal energy, wind and solar energy, ocean energy), the use of hydrogen and other energy sources were not predicted. It should be noted that by 2060, according to forecast data, a gradual increase of electric capacity in the European Union to the value of 1303 GW is planned.

Figure 11 shows forecast data of direct CO_2 emissions from various energy sources for the European Union, according to the "EU Basic Scenario", for the period up to 2060. It shows that by 2060, according to forecast data, the volume of direct CO_2 emissions in general and in the power industry, in particular, will decrease significantly.

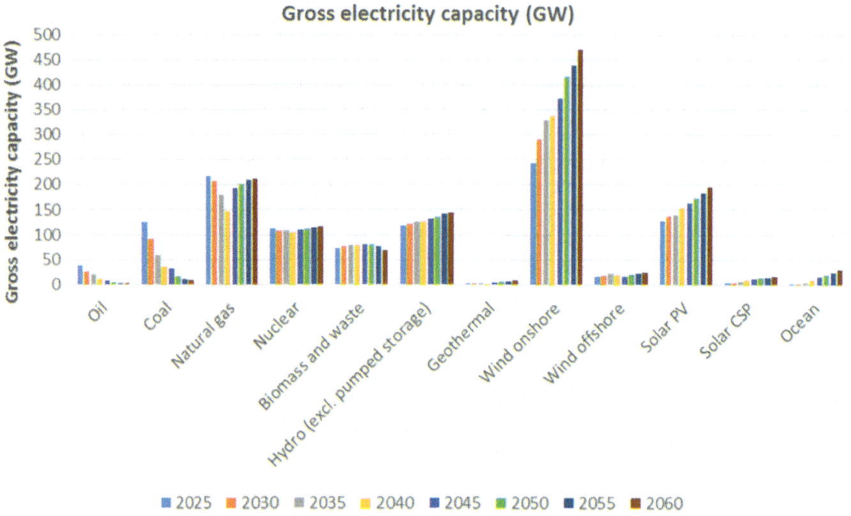

Fig. 10 Forecast data of the growth of electric capacity from various sources for the European Union, according to the "EU Basic Scenario", for the period up to 2060 [24–31]

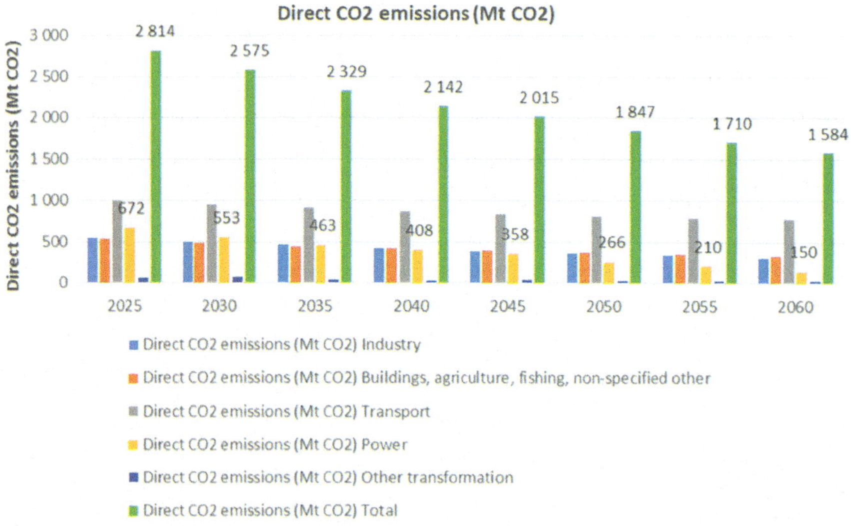

Fig. 11 Forecast data of direct CO_2 emissions from various energy sources for the European Union, according to the "EU Basic Scenario", for the period up to 2060 [24–31]

5 Conclusions

The energy sector occupies a central place in the fight against climate change. The energy sector accounts for two-thirds of the total volume of green house gas emissions, so the energy sector is a central player in ensuring emissions reduction and climate change mitigation. In recent times, promoting sustainable development and combating climate change have become inter-twined aspects of energy planning, analysis and policy development. This study assesses the prospects for using renewable energy sources in the world, Europe and Ukraine in the foreseeable future until 2060, which is considered a means of overcoming the global energy crisis and increasing environmental security in the concepts of sustainable development and "green" energy transition.

The study determined the values of final energy consumption for various energy resources and sources of thermal and electrical energy in Ukraine for 2015–2019. Based on the analysis of the research results, there is an increase in the use of non-traditional and renewable energy sources (biofuel and waste) and a decrease in the consumption of traditional energy resources (coal, natural gas), the use of electricity and thermal energy does not change significantly during the studied period. The structure of final energy consumption by industry in Ukraine in 2019 was evaluated. It was determined that the most significant final energy consumption traditionally falls on the production sector.

The values of global investments in traditional and renewable energy for the period 2019–2021 is estimated. Based on the analysis of the research results, there is a gradual decrease in investments in traditional energy and an increase in financing of non-traditional energy. Estimated forecast data of total primary energy demand from various sources for the European Union, according to the "EU Basic Scenario", for the period up to 2060, as well as forecast (total) values of total primary energy demand. Based on the analysis of the research results, there is a gradual decrease in the overall primary demand for energy from traditional sources (oil, coal, natural gas) and an increase in demand for non-traditional and renewable energy sources. The primary demand for nuclear and hydraulic energy will grow, leading to the insignificant or minimal formation of harmful emissions into the atmosphere. By 2060, according to forecast data, it is planned to reduce the total primary demand for energy from various sources in the European Union to a value of 54,564 PJ.

Estimated forecast data of fuel energy consumption for the generation of electrical and thermal energy from various sources for the European Union, according to the "EU Basic Scenario", for the period up to 2060. It was determined that by 2060, the use of oil, coal, and natural gas in electricity and heat generation would gradually decrease, which will be ensured by the gradual increase in the use of nuclear energy, hydropower, as well as non-traditional and renewable energy sources in heat and electricity generation (biomass and waste, geothermal energy, wind and solar energy, ocean energy), the use of hydrogen and other energy sources was not predicted.

Estimated forecast data for the growth of electricity production from various sources for the European Union, according to the "EU Basic Scenario", for the period up to 2060. Based on the analysis of the research results, there is a gradual decrease in the growth of electricity production from traditional sources and an increase in its generation from non-traditional sources. It was determined that by 2060, the amount of electricity production using oil, coal, and natural gas would gradually decrease, which will be ensured by the gradual increase in electricity production using nuclear energy, hydropower, as well as non-traditional and renewable energy sources in electricity generation (biomass and waste, geothermal energy, wind and solar energy, ocean energy), the use of hydrogen and other energy sources was not predicted. It should be noted that by 2060, according to forecast data, a gradual increase in the volume of electricity production in the European Union is planned to the value of 3519 TWh.

Estimated data of the growth of electric capacity from various sources for the European Union, according to the "EU Basic Scenario", for the period up to 2060. Based on the analysis of the research results, there is a gradual decrease in the amount of electric capacity from traditional energy sources and an increase in the amount of electric capacity from non-traditional and renewable energy sources. It was determined that by 2060, the growth of electric capacity with the use of oil, coal, and natural gas in an electric generation would gradually decrease, which will be ensured by the gradual increase of electric capacity with the use of nuclear energy, hydropower, as well as non-traditional and renewable energy sources in electric generation (biomass and waste, geothermal energy, wind and solar energy, ocean energy), the use of hydrogen and other energy sources was not predicted. It should be noted that by 2060, according to forecast data, a gradual increase of electric capacity in the European Union to 1303 GW is planned.

Estimated forecast data of direct CO_2 emissions from various energy sources for the European Union, according to the "EU Basic Scenario", for the period up to 2060. It was determined that by 2060, according to forecast data, the volumes of direct CO_2 emissions in general and in the energy sector, in particular, will decrease significantly.

The approach proposed in the article makes it possible to reasonably determine the prospects for the use of environmentally safe energy-saving technologies with the use of non-traditional and renewable energy sources in Ukraine in the foreseeable future until 2060, based on the estimated trends in the reform of the energy sector in the world and the European Union, which should be considered as a means of overcoming the global energy crisis and increasing environmental security in the concepts of "green" energy transition.

References

1. UNDP (2018) Transforming our world: the 2030 agenda for sustainable development. https://www.undp.org/ukraine/publications/transforming-our-world-2030-agenda-sustainable-development
2. Support sustainable development and climate action (2019). https://www.un.org/en/our-work/support-sustainable-development-and-climate-action
3. Koval V, Sribna Y, Kaczmarzewski S, Shapovalova A, Stupnytskyi V (2021) Regulatory policy of renewable energy sources in the European national economies. Energy Policy J 24:61–78
4. European Commission (2019) A European green deal. Striving to be the first climate-neutral continent. https://commission.europa.eu/strategy-and-policy/priorities-2019-2024/european-green-deal_en
5. Kreislaufwirtschaftsgesetz (2020) Gesetz zur Förderung der Kreislaufwirtschaft und Sicherung der umweltverträglichen Bewirtschaftung von Abfällen. https://www.bmuv.de/gesetz/kreislaufwirtschaftsgesetz
6. Klaus S (2016) The fourth industrial revolution: what it means, how to respond. World Economic Forum
7. Zero waste (2022). https://zerowaste.org.ua
8. IEA (2022) Renewable energy market update, May 2022. https://www.iea.org/reports/renewable-energy-market-update-may-2022
9. Mihalcova B, Koraus A, Prokopenko O, Hvastova J, Frenákova M, Gallo P, Beata B (2021) Effective management tools for solving the problem of poverty in relation to food waste in context of integrated management of energy. Energies 14:4245. https://doi.org/10.3390/en14144245
10. Net Zero by 2050 (2021). IEA, Paris. https://www.iea.org/reports/net-zero-by-2050
11. Arkhypova LM, Mandryk OM, Moskalchuk NM, Prykhodko MM, Radlovska KO (2021) Renewable energy resources in the system of sustainable development of Carpathian region of Ukraine. J Phys Conf Ser 1781:012010. https://doi.org/10.1088/1742-6596/1781/1/012010
12. Benalcazara P, Komorowska A (2021) Prospects of green hydrogen in Poland: a techno-economic analysis using a Monte Carlo approach. Int J Hydrogen Energy 47(9):5779–5796
13. Chorna M, Filipishyna L, Nord G, Tkachenko S, Velychko K (2019) Formation of analytical support for the management of economic and environment safety of organizations. J Secur Sustain Issues 8:413–422. https://doi.org/10.9770/jssi.2019.8.3(10)
14. Dudek M, Bashynska I, Filyppova S, Yermak S, Cichoń D (2023) Methodology for assessment of inclusive social responsibility of the energy industry enterprises. J Cleaner Prod (in press)
15. Energy technology perspectives (2023). IEA, Paris. https://www.iea.org/events/energy-technology-perspectives-2023
16. Ostapenko O, Olczak P, Koval V, Hren L, Matuszewska D, Postupna O (2022) Application of geoinformation systems for assessment of effective integration of renewable energy technologies in the energy sector of Ukraine. Appl Sci 12:592. https://doi.org/10.3390/app12020592
17. Ostapenko O (2021) Estimation of tendencies of transforming the energy sectors of world, European Union and Ukraine in the perspective to 2050 with using the renewable energy sources in the concept of sustainable development. Social capital: vectors of development of behavioral economics. ACCESS Press Publishing House, Veliko Tarnovo, Bulgaria, pp 99–139
18. Kurbatova T, Sotnyk I, Prokopenko O, Sidortsov R, Tu Y-X (2021) Balancing Ukraine's energy system: challenges under high renewable energy penetration and the COVID-19 pandemic. E3S Web Conf 280:05007. http://doi.org/10.1051/e3sconf/202128005007
19. Malynovska Y, Bashynska I, Cichoń D, Malynovskyy Y, Sala D (2022) Enhancing the activity of employees of the communication department of an energy sector company. Energies 15(13):4701. https://doi.org/10.3390/en15134701
20. Prokopenko O, Chechel A, Sotnyk I, Omelyanenko V, Kurbatova T, Nych T (2021) Improving state support schemes for the sustainable development of renewable energy in Ukraine [Poprawa

programów wsparcia państwa wzakresie zrównoważonego rozwoju energii odnawialnej na Ukrainie]. Polityka Energetyczna 24(1):85–100. http://doi.org/10.33223/epj/134144
21. Sembiyeva L, Serikova M, Satymbekova K, Tulegenova Z, Nurmaganbetova B, Zhagyparova A (2021) Tax audit in innovative development of the energy sector of the economy: global trends. J Water Land Dev 48(1–3):70–80
22. Shpak N, Ohinok S, Kulyniak I, Sroka W, Fedun Y, Ginevičius R, Cygler J (2022) CO_2 emissions and macroeconomic indicators: analysis of the most polluted regions in the world. Energies 15(8):2928. https://doi.org/10.3390/en15082928
23. Sribna Y, Koval V, Olczak P, Bizonych D, Matuszewska D, Shtyrov O (2021) Forecasting solar generation in energy systems to accelerate the implementation of sustainable economic development. Energy Policy J 24:5–28
24. IEA (2022) World energy balance highlights. https://www.iea.org/data-and-statistics/data-product/world-energy-balances-highlights
25. IEA (2022) Energy efficiency indicators highlights. https://www.iea.org/data-and-statistics/data-product/energy-efficiency-indicators-highlights
26. IEA (2022) World energy outlook 2022. https://www.iea.org/data-and-statistics/data-product/world-energy-outlook-2022-free-dataset
27. IEA (2021) World energy investment 2021. https://www.iea.org/reports/world-energy-investment-2021
28. IEA (2021) Global energy review: CO_2 emissions in 2021. https://www.iea.org/data-and-statistics/data-product/global-energy-review-co2-emissions-in-2021
29. IEA (2022) Greenhouse gas emissions from energy highlights. https://www.iea.org/data-and-statistics/data-product/greenhouse-gas-emissions-from-energy-highlights
30. IEA (2022) SDG7 database. https://www.iea.org/data-and-statistics/data-product/sdg7-database
31. IEA (2021) Net zero by 2050 scenario. https://www.iea.org/data-and-statistics/data-product/net-zero-by-2050-scenario
32. IEA (2022) Energy security. Reliable, affordable access to all fuels and energy sources. https://www.iea.org/topics/energy-security
33. IEA (2022) Renewables. https://www.iea.org/fuels-and-technologies/renewables
34. IEA (2022) Total renewable electricity capacity additions, 2001–2027. https://www.iea.org/data-and-statistics/charts/total-renewable-electricity-capacity-additions-2001-2027
35. Sotnyk I, Kurbatova T, Kubatko O, Prokopenko O, Prause G, Kovalenko Y, Trypolska G, Pysmenna U (2021) Energy security assessment of emerging economies under global and local challenges. Energies 14:5860. https://doi.org/10.3390/en14185860
36. IEA (2022) IEA and Ukraine deepen bilateral cooperation with new joint work programme. https://www.iea.org/news/iea-and-ukraine-deepen-bilateral-cooperation-with-new-joint-work-programme
37. IEA (2022) How the European Union can avoid natural gas shortages in 2023. https://www.iea.org/news/how-the-european-union-can-avoid-natural-gas-shortages-in-2023
38. Nitsenko V, Mardani A, Streimikis J, Shkrabak I, Klopov I, Novomlynets O, Podolska O (2018) Criteria for evaluation of efficiency of energy transformation based on renewable energy sources. Montenegrin J Econ 14:237–247. http://doi.org/10.14254/1800-5845/2018.14-4.17
39. Kurbatova T, Sidortsov R (2022) Trash to Hryvnias: the economics of electricity generation from landfill gas in Ukraine. Int J Sustain Energy Plann Manag 33:53–64. https://doi.org/10.5278/ijsepm.6707

Harnessing the Energy of Moving Water to Generate Electricity in Bulgaria

Rositsa Velichkova, Martin Pushkarov, Iskra Simova, Alexander Stanilov, Vladislav Ivanov, Yoana Pavlova, and Angel Alexandrov

Abstract Renewable energy sources such as solar, wind, hydroelectric and geothermal are considered sustainable and environmentally sound alternatives to fossil fuels. The aim of the study is to examine the possibilities for using running water in Bulgaria and to present a test-rig which can be used for utilization of such type of energy. An analysis of the rivers in Bulgaria from the technical catchment point of view was made. The installed and operating hydropower plants in Bulgaria are presented. A developed laboratory test-rig is presented and described in details, which investigates the possibilities of harnessing the energy of rivers (with the use of) hydrokinetic turbines. Results of the theoretical power output of one of the presented turbines were obtained at four different flow velocities in the test-rig. A brief analysis of the social part related to this type of energies was also made.

Keywords Moving water · Harnessing · Hydrokinetic turbine · Energy potential

1 Introduction

Renewable energy sources (wind energy, solar energy, hydroelectric energy, ocean energy, geothermal energy, biomass and biofuels) are alternatives to fossil fuels that contribute to reducing greenhouse gas emissions, diversifying energy supplies and reducing dependence on unreliable and volatile fossil fuel markets, especially oil and gas [1]. EU legislation to promote renewable energy sources, has developed significantly over the last 15 years. In July 2021, in view of the EU's new climate ambitions, the co-legislators were proposed to revise the 40% target by 2030. After the Russian invasion of Ukraine and the subsequent energy crisis, the EU agreed to rapidly reduce its dependence on Russian fossil fuels before 2030, by accelerating the transition to clean energy [2].

R. Velichkova (✉) · M. Pushkarov · I. Simova · A. Stanilov · V. Ivanov · Y. Pavlova · A. Alexandrov
Technical University of Sofia, Sofia, Bulgaria
e-mail: rvelichkova@tu-sofia.bg

V. Koval and P. Olczak (eds.), *Circular Economy for Renewable Energy*, Green Energy and Technology, https://doi.org/10.1007/978-3-031-30800-0_8

1.1 Biomass

Biomass is considered one of the best examples of non-conventional energy source. It is also considered a suitable alternative to fossil fuels for the production of electricity. It is known that the general name biomass usually refers to raw materials obtained from wood waste, waste from agriculture and the food industry, as well as plants and trees, grown for the purpose of being used as a raw material in the production of energy. Biomass also includes sediments obtained during wastewater treatment, as well as manure [3]. The main advantages of biomass are the fact that it is a renewable energy source and its wide availability, and in large quantities. Its advantages are its relatively lower price, the small investment associated with the creation and operation of such a station, the possibility of usefully utilizing part of the accumulated waste. Promoting the use of biomass also plays an important role in reaching the European Commission's climate change targets. According to experts, this is probably due to the fact that so far for Europe, energy produced from biomass constitutes about 2/3 of all energy produced from renewable energy sources [3].

1.2 Solar

Solar power generation systems, convert sunlight into electricity. In order to carry out this transformation of energy from solar to electricity, photovoltaics do not need direct sunlight, but simply daylight. The higher the light intensity, the greater the amount of electricity produced.

1.3 Wind Power

Wind energy is converted into electricity with wind turbines. Sailing boats and ships are a good example of using the mechanical wind energy. The production of electricity is the most important application of wind power energy in the present moment. The mechanical energy of the wind is captured by wind turbines that drive an electric generator for electricity production. Experts are on opinion that wind is one of the most promising energy sources. Today, wind energy is the fastest-growing sector, with an increase of 22%. The reasons are that the price of electricity produced by wind is, in many cases, lower than the energy produced by the sun because wind facilities have lower installation and maintenance costs [4].

1.4 Hydroelectric Power

Water energy is harnessed by various machines that use the power of the moving water. This energy is utilized either directly as mechanical energy or is converted into electrical energy. The oldest examples of use of hydropower are water mills. The main sources can be divided into ocean and river hydropower sources.

Ocean energy sources: useful energy from ocean waters can be harnessed by the waves in the ocean. Water collects in certain places. As the water flows in or out of these natural or artificially created cavities, it can drive water turbines and generate electricity [5].

1.5 River Energy Sources

The energy of moving water in rivers can be obtained by means of water turbines which are connected to generators for electricity production. To ensure constant production of electricity, special hydrotechnical facilities are most often used (dams) [6, 7].

Experiments show that it is better to produce energy from sea or ocean waves not near the coast, where the waves are already weakened, but in the open sea or in the coastal zone.

1.6 Geothermal

"Geothermal"—the term comes from the Greek word "geo"—earth and "therma"—heat, which in the general meaning of the word means—thermal energy from the Earth.

However, heat in the deep and shallow parts of Earth comes from two distinctly different sources. The first source is deep heat energy generated within the Earth itself by the radioactive decay of unstable isotopes and friction. The second source is shallow heat energy from the sun that is stored in the ground.

The practical use of geothermal energy depends on the localization of the source, the flow rate, its temperature, its proximity to the users, the climate conditions and the built infrastructure [8].

Tables 1, 2, 3, 4 and 5 represent the renewable energy percentage distribution worldwide [9].

The data organised in Table 1 shows almost 10% increase in hydropower energy production in Asia, while Europe and North America show a 5% decrease for the period analysed. The rest of the regions have no change throughout the years 2009–2021.

Table 1 Hydropower energy percentage distribution by region for the period 2009–2021

	Africa	Asia	Central America and Caribbean	Eurasia	Europe	Middle East	North America	South America	Oceania
2009	3	35	0	7	21	1	19	13	1
2010	3	35	1	7	20	1	18	13	1
2011	2	37	1	7	20	1	18	13	1
2012	3	38	1	7	19	1	17	13	1
2013	2	40	0	7	19	1	17	13	1
2014	2	41	1	7	18	1	16	13	1
2015	2	41	1	7	18	1	16	13	1
2016	3	41	1	7	17	1	16	13	1
2017	3	42	1	7	17	1	15	13	1
2018	3	42	1	7	17	1	15	13	1
2019	3	42	1	7	17	1	15	13	1
2020	3	43	0	7	17	1	15	13	1
2021	3	44	1	7	16	1	14	13	1

Table 2 Wind energy percentage distribution by region

	Africa	Asia	Central America and Caribbean	Eurasia	Europe	Middle East	North America	South America	Oceania
2015	1	39	0	1	34	0	21	3	1
2016	1	40	1	0	33	0	21	3	1
2017	1	40	1	0	34	0	20	3	1
2018	1	41	1	0	32	0	20	4	1
2019	1	42	1	0	32	0	20	3	1
2020	1	46	1	0	29	0	19	3	1
2021	1	47	1	0	27	0	19	4	1

Table 3 Solar energy percentage distribution by region

	Africa	Asia	Central America and Caribbean	Eurasia	Europe	Middle East	North America	South America	Oceania
2015	1	40	0	0	44	0	12	3	0
2016	1	47	0	0	35	1	13	1	2
2017	1	54	1	0	28	1	12	1	2
2018	2	56	1	0	25	1	12	1	2
2019	2	56	1	0	24	1	12	2	2
2020	2	57	1	0	23	1	12	2	2
2021	1	57	1	1	22	1	13	2	2

Table 4 Bioenergy percentage distribution by region

	Africa	Asia	Central America and Caribbean	Eurasia	Europe	Middle East	North America	South America	Oceania
2015	1	25	2	2	36	0	17	16	1
2016	1	29	2	2	34	0	16	15	1
2017	2	30	2	2	33	0	15	15	1
2018	1	32	2	2	34	0	14	14	1
2019	1	34	2	2	33	0	13	14	1
2020	1	37	2	2	31	0	12	14	1
2021	1	40	2	2	29	0	12	13	1

Table 5 Geothermal energy percentage distribution by region

	Africa	Asia	Central America and Caribbean	Eurasia	Europe	Middle East	North America	South America	Oceania
2015	5	33	5	6	13	0	29	0	9
2016	6	33	5	7	13	0	28	0	8
2017	6	33	5	9	12	0	27	0	8
2018	5	33	5	10	12	0	26	1	8
2019	5	33	5	12	12	0	26	0	7
2020	6	32	5	12	12	0	25	0	8
2021	5	30	5	11	11	0	31	0	7

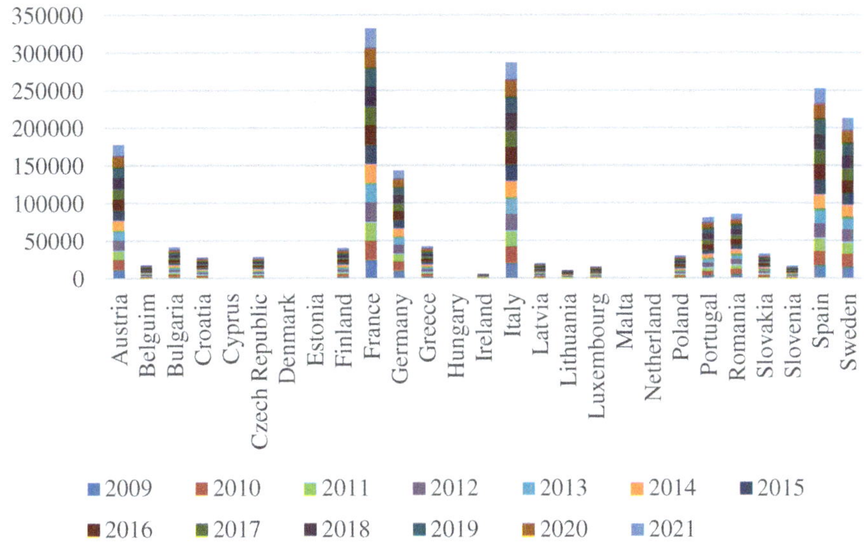

Fig. 1 Hydropower energy in EU for 2009–2021

According to Table 2, there is a 7% decrease in wind energy production in Europe and a 3% decrease in North America, while Asia shows increase of 8% for 2015–2021.

Similarly to the hydropower and wind energy production analyses, the solar energy production in Asia increased significantly for the period considered, and decreased with 22% in Europe (Table 3). The rest of the regions do not show notable change in this part of renewable energy production.

The same trend is observed in the analyses of bioenergy (Table 4) and geothermal energy production (Table 5) by region.

On Figs. 1, 2, 3 and 4 the distribution of renewable energy in the European Union for a period 2009–2021 is shown [10].

The purpose of this chapter is to show how water energy can be used in Bulgaria. Existing hydroelectric plants are shown and explained. It is discussed how hydrokinetic turbines can also be used.

2 Rivers in Bulgaria

There are two river drainage basins in Bulgaria: Black Sea and Aegean Sea. The Black Sea basin includes the rivers that flow directly into the Black Sea or via the Danube River. This water basin occupies 57% of the country's territory. Of the rivers that belong to the Black Sea basin and flow through the Danube, the longest and with the largest catchment basin is the Iskar river—368 km. The White Sea

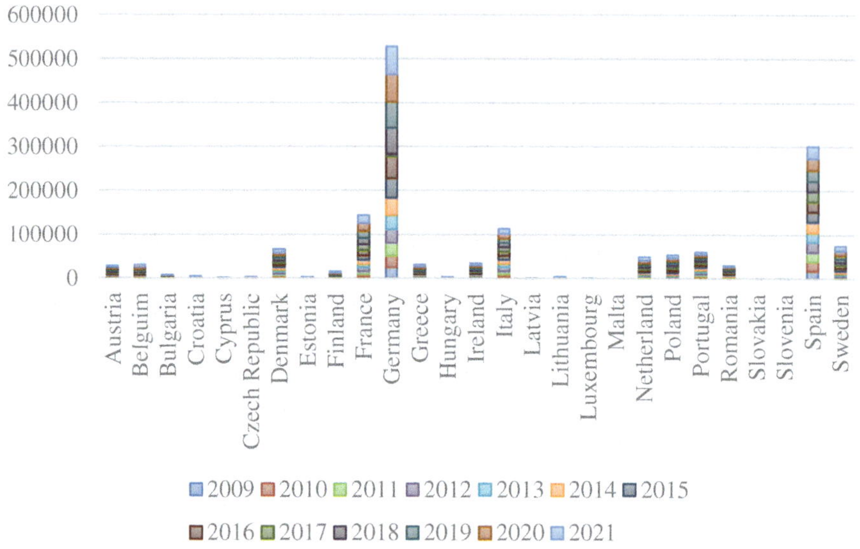

Fig. 2 Wind energy in EU for 2009–2021 period

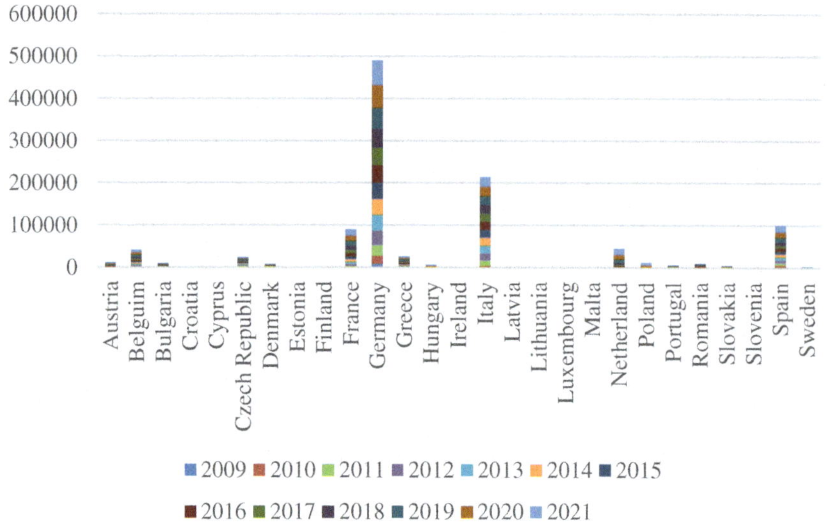

Fig. 3 Solar energy in EU for 2009–2021 period

basin covers 43% of the country's territory. Most of the water flows from Balkan Mounties, Rhodope Mountain and Rila Mountain. The largest of the rivers that flow into the White Sea Basin is the Maritsa River. This basin also includes Tundzha, Arda, Struma and Mesta [11, 12]. The classification of rivers by length reveals the

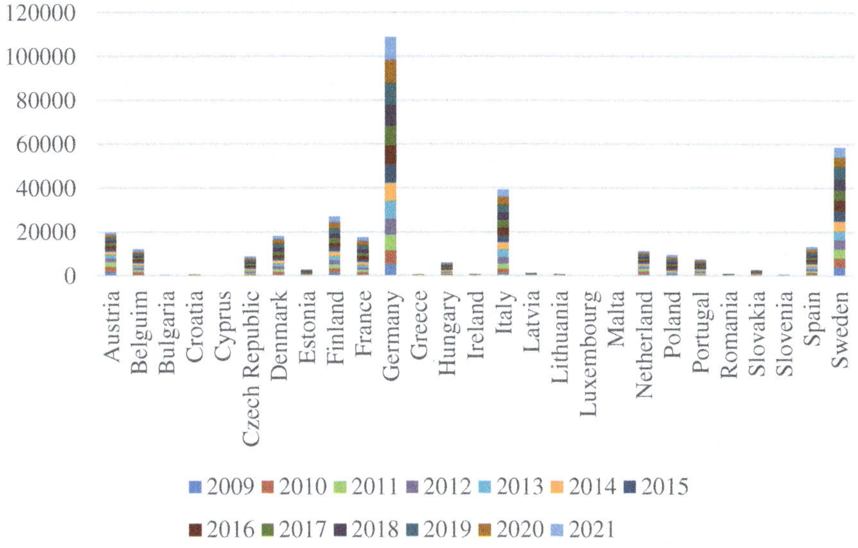

Fig. 4 Bioenergy in EU for 2009–2021 period

following features. Class "very large in length" includes nine rivers: Iskar (368.0 km), Tundzha (349.5 km), Maritsa (321.6 km), Osam (314 km), Struma (290.0 km), Yantra (286.0 km), Kamchiya (245.0 km), Arda (241.0 km) and Luda Kamchiya (200.9 km) (Table 2). Large rivers in length are Rusenski Lom (196.9 km), Vit (188.6 km), Rositsa (164.3 km), Topolnitsa (154.8 km), Veleka (147.0 km), Sazliyka (145.4 km), Ogosta (144.1 km), Beli Lom (140.7 km), Skut (196.9 km), Cherni Lom (130.3 km), Mesta (125.9 km), Sukha–to the border (125.8 km), Provadiyska (119.0 km), Strumeshnica (114.0 km), Rezovska (112.0 km), Stryama (110.1 km), Kanagiol–to the border (109.6 km), Tsaratsar (108.0 km) and Senkovitsa (101.6 km) [12].

Medium-length rivers are 62—the rivers Varbitsa (98.1 km), Dospat (96.2 km), Dobrotinsko (92.9 km), Lom river (92.5 km), Harmanliyska (91.9 km), Lefeja River (91.8 km), Hersovska River (90.8 km) and others. They are evenly distributed between the northern and southern parts of the country. Geographically, this class of rivers stands out in the northwestern parts of the country with the watercourse of Tsibritsa, Lom, Vidbol, Topolovets, Botunya, etc., Northeastern Bulgaria—with the rivers Dobrotinsko, Topchiyska, Golyam Kachamak, Khrsovska, Neykovsko dere, Dobrichka, Karaman etc., Rhodopes—with the Dospat, Asenitsa, Chepinska, Stara (Peshterska) rivers, the Burgas lowland—with the watercourses of the Fakiyska, Gospodarevska, Sredetska, Rusokastrenska and Upper Thracian Plain—Mochuritsa, Kalnitsa, Popovska reka, Pyasčnik, Ovcharitsa, Sokolitsa (Table 6) [11].

Medium in length are 222 watercourses. This class includes 40% of the rivers in Bulgaria—the rivers Archar (59.4 km), Dryanovska (59.3 km), Krumovitsa (58.5 km), Omurovska (57.5 km), Malki Lom (57.1 km) etc. There are no geographical regularities in the distribution of medium-length rivers.

Table 6 Distribution (number) of rivers by length

Class	Length (km)	Black Sea hydro-geographical region		White Sea hydro-geographical region	Total for the country
		Danube region	Black Sea region		
Very small	To 10	18	5	19	42
Small	10.1–20	47	42	97	186
Medium	20.1–50	68	41	113	222
Medium-sized	50.1–100	27	8	27	62
Large	100.1–200	11	3	6	20
Very large	Over 200	3	2	4	9

Small rivers in length—186, as well as very small ones are in mountainous regions. The very small rivers in length are significantly larger number than the one indicated, therefore the class has a higher relative share of the results obtained, and probably almost equal to or greater than the number of rivers in the small and medium classes (Table 7).

The classification of rivers by area of the river basin is similar to the classification of rivers by length—there are the fewest rivers with a very large catchment area, the largest number of rivers in the small and medium classes.

The Maritsa (21,084 km^2) and Struma (10,797 km^2) rivers form the largest catchment areas. The river basin of Maritsa is about 1/5 (18.9%) of the territory of the country, and that of the Struma river—9.7%. The rivers that have large river basins are: the Iskar River (8642.2 km^2), the Tundzha River (7883.6 km^2), the Yantra River (7861.6 km^2), the Kamchia River (5357.6 km^2), the Arda River (5201.2 km^2), Sazliyka River (3293.0 km^2), Vit River (3225.0 km^2), Ogosta River (3157.1 km^2), Rusenski Lom River (2946.9 km^2), Osam River (2824.1 km^2), Mesta River (2767.0 km^2), Suha Reka River (2403.9 km^2), Rositsa River (2264.6 km^2)

Table 7 Distribution (in number) of rivers by catchment area

Class	Area (km^2)	Black Sea hydro-geographical region		White Sea hydro-geographical region	Total for the country
		Danube region	Black Sea region		
Very small	To 20	18	2	22	42
Small	20.1–100	64	60	128	252
Medium	100.1–500	59	31	87	177
Medium-sized	500.1–2000	26	6	23	55
Large	2000.1–10,000	8	2	4	14
Very large	Over 10,000	–	–	2	2

and Provadiyska River (2131.8 km^2). Almost all rivers of this class are in Northern Bulgaria. 10% of the country's river basins are medium large in area. They are developed throughout the country and especially in the Ludogorsk-Dobrudzha region (Tsaratsar, Kanagyol, etc.), Northwestern Bulgaria (the river basins of the rivers Tsibritsa, Lom, Botunya, etc.), the eastern part of the country (the rivers of the Veleka, Sredetska, Fakiyska, etc.), Rhodopes (watersheds of the Vacha, Chepelarska, etc.), Tundzhan lowland (river basins of the rivers Mochurica, Sinapovska, Kalnitsa, etc.), Eastern Rhodopes (watershed of the rivers Varbitsa, Krumovitsa) and the Upper Thracian Plain (the basins of the Topolnitsa River, the Stryama River, the Harmanliyska River, etc.) Average in area are 178 river basins. They form the second largest river class in the grouping. They are distributed almost evenly between Northern and Southern Bulgaria. The small catchments are 46% of all in the country and are mostly mountainous.

Exceptions are river basins from the Tundzhan lowland (on the Kushudere River and on the Slivetovska River—tributaries of the Popovska River). First-order rivers in mountainous regions are usually very small in area. Only the Ludnya River (a tributary of the Tsaratsar River) in the northeastern parts is flat. Table 8 shows the largest rivers in Bulgaria with their length, catchment area and flow rate [12].

3 Hydropower Installed Capacities

Three quarters of the Earth's surface is occupied by oceans, seas, lakes and rivers. Water power has been used since ancient times, with the first water engines being used in irrigation systems to drive mills, etc. The first hydroelectric power plant (HPP) was built and put into operation on the Iskar River in 1900 based on a project of French engineers. Its initial power capacity was 736 kW (0.736 MW). Its purpose was to provide lighting in the capital of Bulgaria, the city of Sofia, and to power the tram traffic. In the period from 1900 to 1918 5 small hydroelectric power stations with a total installed capacity of 4300 kW (4.30 MW) were built in Bulgaria. In 1940 the state took over the electrification of Bulgaria, adopting 2 years later in 1942 the first electrification plan for Bulgaria for the period up to 1960, planning the construction of a 110 kV power ring. Unfortunately, this plan is not implemented, due to the fact that almost all construction works, as well as the supply of materials and equipment, are carried out by foreign companies and specialists [13, 14].

After September 9th, 1944, (September 9th coup) and the socialist restructuring/reconstruction, as a result of the rapid pace of development of Bulgarian industry and agriculture, the need for water for irrigation of cultivated crops and electricity production from water sources, as well as additional water supply in the different regions in Bulgaria increased. During this period, there was a shortage of specialists with experience in the field and several HPPs similar to the predecessors were built, such as HPPs "Koinare", "Mezdra", "Ticha", "Petrovo", etc. During the construction of these HPPs, our specialists gained experience, and meanwhile new Bulgarian specialists were trained [14].

Table 8 Largest Bulgarian rivers by length, catchment area and flow rate

№	Name	Length (km)	Watershed (Drainage basin) (km^2)	Drainage (m^3/s)
1	Iskar	368.0 352.3	8646.2 8617.1	54.5
2	Tundzha	349.5 328.4	7884.0 7890.7	39.7
3	Maritsa	321.6 309.0	21,084.0 21,271.9	108.0
4	Osum	314.0 204.6	2824.0 2838	16.40
5	Struma	290.0 275.1	10,797.0 8036.3	76.167
6	Yantra	285.5 222.0	7862.0 7862.4	36.8
7	Kamchiya	244.5 190.7	5357.6 5363.0	24.476
8	Arda	241.3 229.1	5201.0 5213.1	15.213
9	Luda Kamchiya	200.9 180.2	1611.9 1610.0	9.681
10	Golyama Kamchiya	198.5	2997.0	11.40
11	Rusenski Lom	196.9 155.5	2947.0 2985.0	5.60
12	Vit	188.6 167.5	3225.0 3227.6	8.67
13	Rositsa	164.3 148.5	2265.0 2260.0	10.60
14	Topolnitsa	154.8 135.6	1789.0 1780.0	10.00
15	Veleka	123.0 107.9	955.0 791.9	7.694
16	Sazliyka	145.4 121.7	3293.0 3366.0	18.00
17	Ogosta	141.1 142.7	3157.0 4282.2	23.554
18	Beli Lom	140.7 122.8	1279.0 1310.0	2.09
19	Skat	134.0 139.8	1074.0 1070.0	1.7
20	Cherni Lom	130.3 114.2	1549.0 1550.0	
21	Mesta	125.9 126.1	2767.0 2785.5	29.85

(continued)

Table 8 (continued)

№	Name	Length (km)	Watershed (Drainage basin) (km^2)	Drainage (m^3/s)
22	Suhata reka	125.8 148.4	2404.0 2421.3	0.69
23	Provadiya	119.0 120.9	2132.0 2145.3	3.10
24	Vacha	111.5 103.7	1645.0 1680.0	22.0
25	Stryama	110.1 98.7	1394.0 1490.0	10.05
26	Kanagyol	109.6 85.4	1745.0 1785.5	0.086
27	Senkovets	101.6 121.7	553.0 3366.0	0.166
28	Varbitsa	98.1 99.2	1203.0 1207.6	
29	Lom	93.0 102.9	1240.0 1159.6	7.39

In 1946 started the construction of the first Bulgarian dam entirely by Bulgarian specialists. The name was "Tashboaz" dam (the name comes from the Tashboaz area), which was later renamed "Vasil Kolarov" and currently bears the name "Golyam Beglik", as it was completed and put into operation in 1951 [14].

Bulgarian hydropower is growing stronger with the development of Bulgaria in all spheres, and in 1947 the construction of the "Petrohan" cascade began (Fig. 5), which was completed and put into operation in 1956. During these nine years, the three HPPs of the cascade "Petrokhan" (7.7 MW), "Berzia" (5.6 MW) and "Klisura" (3.5 MW) were built with a total installed capacity of 16.8 MW [14].

For the period up to 1948, a total of 47 hydropower plants with a total capacity of 46.5 MW were built in Bulgaria. At that time, the largest hydroelectric plant in Bulgaria was "Vacha" with an installed capacity of 7.0 MW. The "Beli Iskar" dam and the "Beli Iskar" low dam hydroelectric power plant with an installed capacity of 16.8 MW establish/found the Iskar cascade (Fig. 6). In 1951, the construction of the largest hydroelectric plant along the Iskar River began. This cascade includes the "Iskar" Dam, the "Pancharevo" Dam, the "Kokalyane" Dam and the two HPPs "Pasarel" and "Kokalyane" with a power of 28 MW and 22 MW respectively [15].

The cascade entered service in 1956. In 1957, the construction of the "Studen Kladenetz" Dam and the "Studen Kladenetz" Hydroelectric Power Plant with an installed capacity of 60 MW was completed, and to date it has a capacity of 85.5 MW (Fig. 7). At the end of 1964, the "Ivaylovgrad" and "Kardjali" dams were completed, as well as the adjacent hydroelectric power plants. The capacities of the hydroelectric power plants are as follows: for "Ivaylovgrad" it is 120 MW and for "Kardjali" it is 124 MW [14].

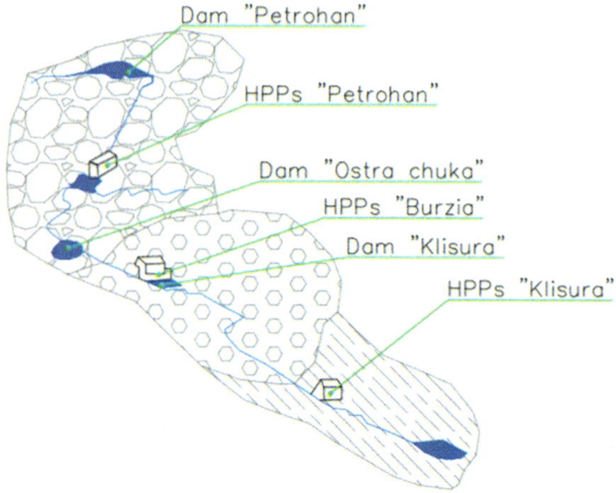

Fig. 5 Petrohan cascade

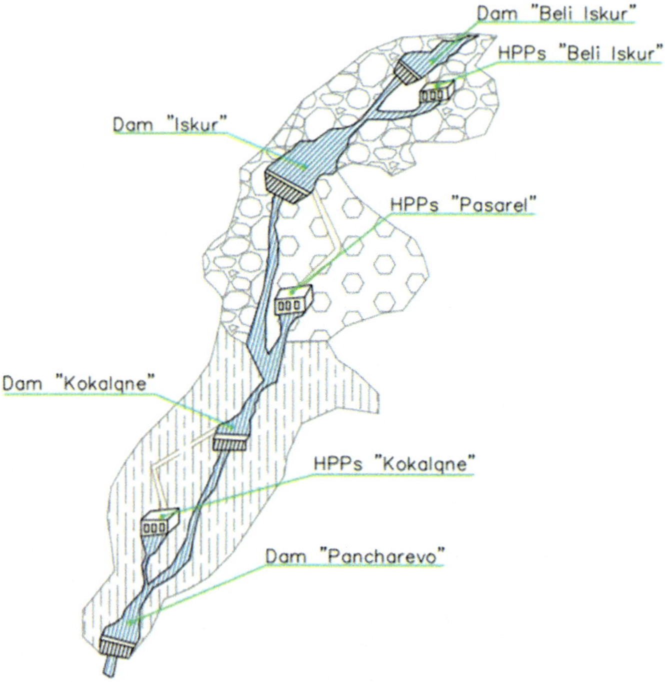

Fig. 6 Iskar cascade

Fig. 7 Arda cascade

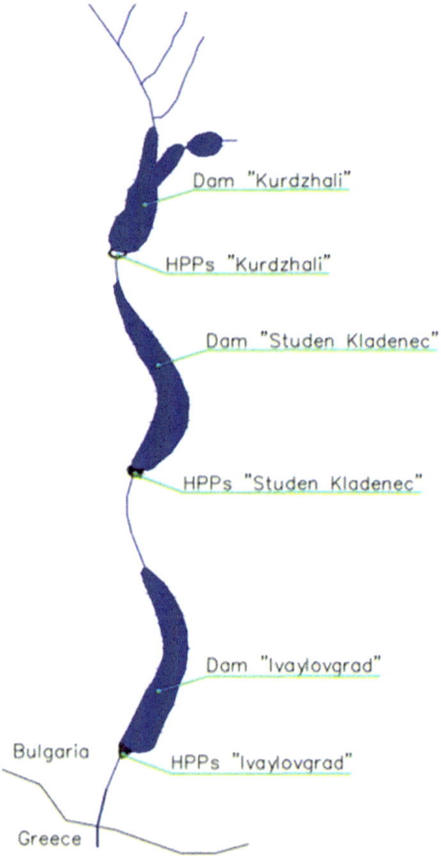

In 1961, the construction of the "Topolnitsa" Dam and the 9 MW "Topolnitsa" Hydroelectric Power Plant adjacent to it was completed. In 1962, the construction of another mega hydropower cascade "Dospat–Vacha" began (Fig. 8).

The first "Vasil Kolarov" dam, or currently the "Golyam Beglik" dam, marks the beginning of the largest hydropower cascade in Bulgaria to date, "Batashki hydroelectric road", with its construction and entry into service in 1969 (Fig. 9). It includes the dams "Batak", "Beglika" and "Toshkov chark", as well as "Batak" 46 MW, "Peshtera" 136 MW and "Aleko" 71.4 MW Hydroelectric Power Plants. When the cascade was completed, it was the most intricate hydrotechnical complex not only in Bulgaria but also in all of Europe, including 93 km of canals, 72 km of tunnels, 4.65 km of dukers and over 170 mountain catchments.

The largest construction projects in Bulgaria's hydropower system began somewhere from 1966 to 1975 with the "Belmeken–Sestrimo" cascade, about 5106 m^3 of external excavations, 687,103 m^3 of concrete work, 673,103 m^3 of formwork, over 5.106 m^3 bulk works and others (Fig. 10). The three HPPs along the cascade have capacities as follows: PHES "Belmeken" 375 MW (commissioned in 1973, its

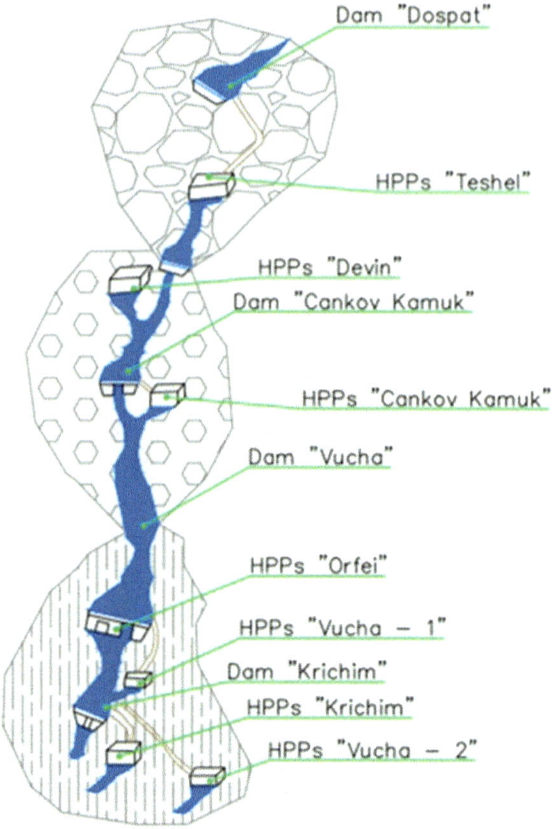

Fig. 8 Dospat–Vacha cascade

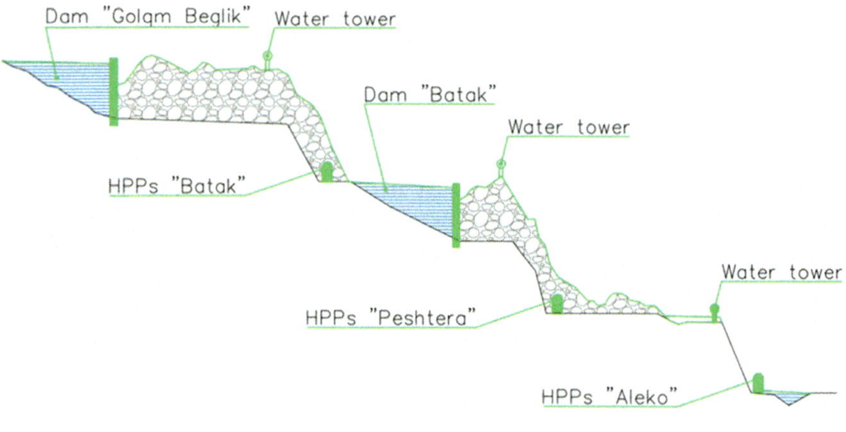

Fig. 9 Batashki vodnosilov pat cascade

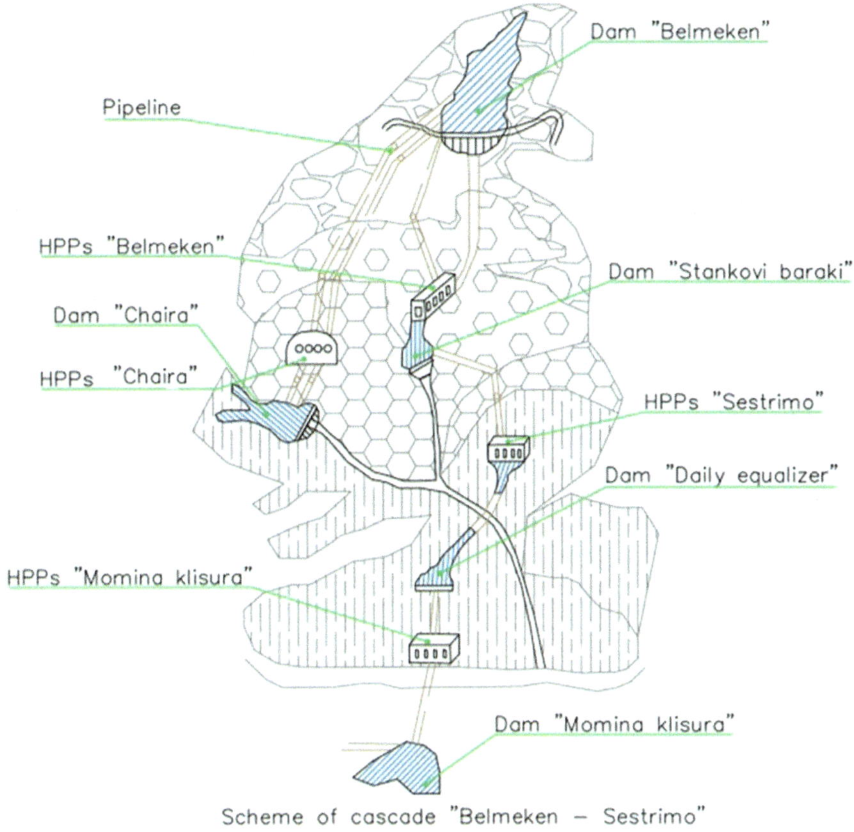

Fig. 10 Belmeken–Sestrimo cascade

power in pumping mode is 104 MW), HPP "Sestrimo" 260 MW and HPP "Momina Klisura" 120 MW [14].

In 1975, the 4th stage of the "Dospat–Vacha" cascade, PAVEC "Orpheus" with a total installed capacity of 160 MW in turbine mode and 43 MW in pump mode, was put into operation.

In order for Bulgaria's energy system to have a balance between consumption and production, it is necessary to design a PHES with a large capacity. In 1970, the exploration and design of the facility began, and the construction of Bulgaria's first classic PHES "Chaira" began in 1980. PHES is a unique hydropower facility not only in Bulgaria but also in Europe (Fig. 10).

"Chaira" has 4 pump-turbine hydro units with a total capacity of 864 MW or 216 MW per unit in turbine mode. When "Chaira" works in pumping mode, the hydro units realize 788 MW power. PHES "Chaira" is an underground power station, with the machine room located 350 m underground in a cavern measuring 22.5 × 11.5 m with a maximum height of 43 m. Reversible machines or so-called hydro units are of the Francis type (the turbines), and their maximum flow rate in turbine mode (the amount of water passing through one turbine) is 36.0 m^3/s. In pump mode the flow rate through the wheel (maximum) is 29.3 m^3/s. In 1995, hydrounit № 1 and hydrounit № 2 came into operation with the entire equipment delivered from Japan, and in particular by the company "Toshiba". The second stage of "Chaira" was in the period 1996–1999, when the pressure pipeline № 2, water tower and main transformer № 2, and hydro units № 3 and № 4 were built. The equipment in the second stage was built under the Japanese licenses of "Toshiba", but this time in Bulgaria and in Bulgarian enterprises [15].

The Rila Cascade is a hydroelectric cascade in western Bulgaria in the western part of the Rila Mountains, consisting of a series of dams.

The cascade is the first large hydropower project in Bulgaria, and its construction began in 1924. With a relatively small catchment area, it has a very high total geodesic drop—2235.5 m—and includes the highest dams in the country (Fig. 11).

On Fig. 12 is given scheme of creating a pad and collecting the water in dams in Bulgaria.

So far, the hydropower industry of Bulgaria has been briefly described, as it has its rise and its peak, to date, many small hydropower plants with small power have been implemented, many reconstructions, rationalizations and modernizations have been

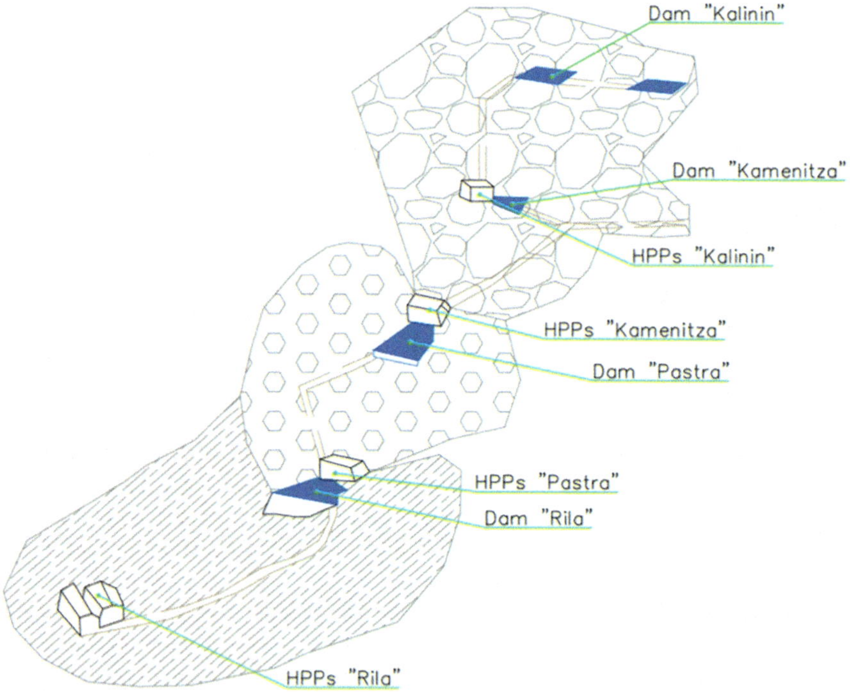

Fig. 11 Rila cascade

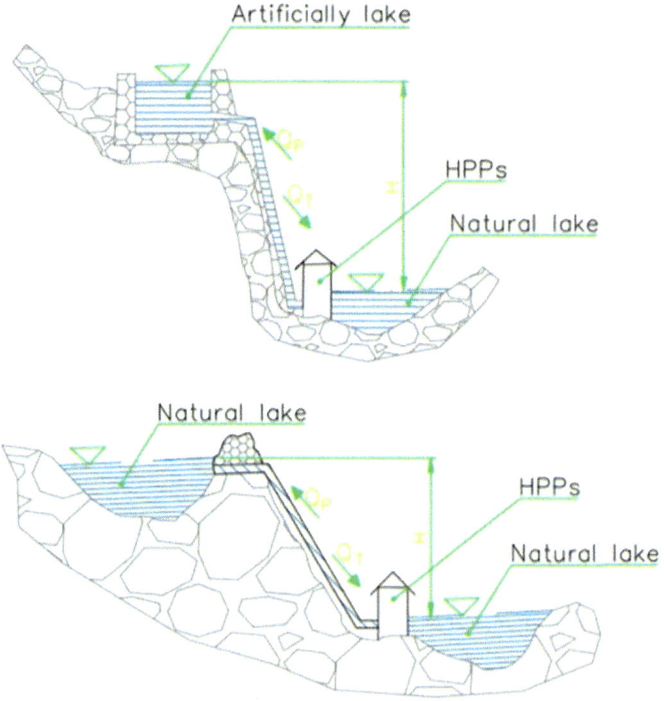

Fig. 12 Scheme of creating a pad and collecting the water

carried out in many of Bulgaria's hydropower plants. In this chapter, the significant hydropower projects of Bulgaria are presented, numerous data are presented on the realized power of hydropower plants, a sample of the types of turbines used in our power plants is presented, as well as the maximum power realized by hydropower in Bulgaria. Although a small country (in terms of territory), Bulgaria was and still is one of the countries with the most well-developed "Energy", in general. On its territory, there are built nuclear power plants, thermal power plants (lignite coal), thermal power plants (gas), thermal power plants (hard coal), hydroelectric power plants, hydroelectric power plants, thermal power plants, thermal power plants and bioelectric power plants.

Bulgaria's energy system has an installed capacity of 13 MW. The development of water engines is directly related to the increase in the energy needs of mankind. Nowadays, methods, machines and equipment based on the use of the energy of water located above sea level or rivers and lakes find significant practical application. Technologies utilizing wave, hydrokinetic and ocean power are also developing very seriously at the present time. Water energy resources are derivatives of solar energy and belong to the so-called renewable energy resources or sources.

Advantages of HPP over NPP and CHP:

1. Renewable energy is used;
2. High efficiency of the hydroelectric plant (total efficiency of the hydroelectric plant approximately 88–90%) compared to the other two plants;
3. Low cost of the energy produced by the hydroelectric plant (less service person, but low costs, automated systems, etc.);
4. High maneuverability of hydro units;
5. High reliability of all machines and equipment in the HPP;
6. Low operational and environmental problems;

Disadvantages of HPP:

1. High initial capital value;
2. Flooding of arable lands;
3. Increasing impact on the environment;
4. Relocation of entire settlements whose area will be flooded;

Hydroelectric plant:

The creation of a pad (geodesic head) of water and the conversion of its energy into electricity is carried out by a complex system of facilities called a hydro node. The most common schemes for creating a geodesic pressure are:

1. Dam;
2. Derivation scheme;
3. Mixed scheme;

Pumped hydroelectric energy storage (PHES) is a complex of machines and equipment not only for generating, but also for accumulating energy. The hydraulic machines installed in the plant must ensure its efficient operation in pump and turbine mode.

Classification of hydroelectric power plants according to their scheme:

1. According to the water regime:
   - HPP of running water—the natural regime of the water source is used;
   - Hydroelectric power plants on equalized waters—have a reservoir in which the volume of water can be regulated in a certain time interval;
   - Hydroelectric cascade—several plants connected through a common water regime;
2. According to head;
   - Low head—with head up to 30–40 m;
   - Medium head—with head of 40–150 m;
   - High head—with head over 150 m;
3. According to the nature of work in the energy system;
   - Basic—work continuously and provide basic needs of the energy system;
   - Peak—work during the hours of peak load of the energy system;

National Electric Company EAD is the largest producer of electricity from water resources in Bulgaria. The company is state-owned and owns 31 hydroelectric power plants (28 HPP and 3 PAVEC) with a total installed capacity of 2737 MW in turbine mode and 931 MW in pump mode. With its capacities, National Electric Company EAD actively participates in the management and regulation of the energy system of our country. HPP and PHES are a major contributor in Bulgaria's electricity system with their fast-maneuvering capacities and wide coverage range [14, 15]. This data is given at Table 9.

Table 10 illustrates the construction of hydroelectric power plants in Bulgaria over the years. Now, approximately 3.2 MW of hydropower plants have been installed on the territory of Bulgaria, and the total number of plants is about 290 [15].

According to the head of the hydropower plants in Bulgaria, the plants are divided into Micro, Mini and Small, as shown in Table 11.

For the territory of Bulgaria, it was accepted that small hydropower plants have a capacity of up to 2 MW. To date, the classification of HPP is the same as in the whole EU and is presented in Table 12 [16].

The studies for small hydropower plants in Bulgaria were carried out by "Energoproekt" already in the period from 1982–1986, showing that in Bulgaria there is a technical potential for the construction of approximately 700 new hydropower plants with small power, as their total installed capacity estimated to be around 250 MW [13, 16].

From the various types of water turbines used in Bulgarian hydropower plants presented in Table 13 of them were manufactured and installed by Bulgarian enterprises.

4 Flowing Water Utilization via Hydrokinetic Turbine

Hydrokinetic turbines can be said to be analogous to wind turbines because aerodynamics and hydrodynamics share common principles. On the other hand, regardless of the general principles, there are also significant differences between wind and hydrokinetic converters of kinetic energy of fluid flows. Some of the significant differences are:

- Water is a practically incompressible fluid and air is a compressible fluid;
- Water currents are established and with completely predictable directions, while winds can blow from all directions;
- The density of water is 800 times greater than the density of air.

Due to the differences listed above and other reasons, the approaches to the development and design of hydrokinetic systems for converting the kinetic energy of fluid flows into useful energy, including electrical energy, are also different.

The system approach to the design of hydrokinetic power plants requires an optimal choice of the entire scheme of the power plant: from the primary conversion of water energy to the last step—the production of electric current. The design of

Table 9 The state-owned HPPs and PAVEC in Bulgaria as of 31.12.2021

Cascade	Power plant name	Power plant type	Installed capacity		Ownership
			Turbine mode (MW)	Pump mode (MW)	
Belmeken–Sestrimo–Chaira	Belmeken	***PHES***	375	104	State
	Sestrimo	HPP	240		
	Momina Klisura		120		
	Chaira	***PHES***	864	784	
The Batash hydroelectric road	Batak	HPP	46		State
	Peshera		136		
	Aleko		71.4		
Dospat–Vacha	Teshel	HPP	60		State
	Devin		88		
	Tsankov Kamak		85		
	SHP Tsankov Kamak		1.3		
	Orpheus	***PHES***	160	43	
	Krichim	HPP	80		
	Vacha 1		14		
	Vacha 2		7		
Arda	Kardzali	HPP	124		State
	Studen Kladenets		81.3		
	Ivaylovgrad		120		
Lakatnitsa	Beli Iskar	HPP	16.8		State
	Mala Zarkwa		7.7		
Small hydro power	Simeonovo	*SHP*	6.67		State
	Boyana		1.24		
	Lukovit		0.69		
	Troyan 1		0.22		
	Vidima		3.3		
	Rositsa 1		7.5		
	Toplnitsa		9.4		
	Ustovo		0.64		
	Byal izvor		1.85		
	Levski		3.2		

(continued)

Table 9 (continued)

Cascade	Power plant name	Power plant type	Installed capacity		Ownership
			Turbine mode (MW)	Pump mode (MW)	
	Tyzha		4.9		
Danube	Kozloduy	HPP	5		State
Strumsko	Strumsko	*SHP*	0.15		State

Table 10 Number of HPPs built and put into operation over the years

Type of HPP	Year	Number of HPPs put in operation
HPP with a capacity of up to 1 MW	1912 until 1966	22
	1993 until 2019	57
HPP with a capacity of more than 1 MW	1931 until 1992	42
	2003 until 2020	32
SHP with a capacity of up to 1 MW	1941 until 1998	5
	2000 until 2019	53
SHP with a capacity of more than 1 MW	2000 until 2021	37

Table 11 Division of HPPs according to their head

HPP	HEAD (m)		
	Low	Medium	High
Micro	< 15	15–50	> 50
Mini	< 20	20–100	> 100
Small	< 25	25–130	> 130

Table 12 Classification of hydropower plants according to their power capacity

Country	HPP power capacity (kW)		
	Micro	Mini	Low
Russia	< 100	100–1000	1000–30,000
USA	< 100	–	1000–30,000
China	–	< 500	500–25,000
India	< 100	100–1000	1000–15,000
France	5–5000 kW		
Brazil	< 100	100–1000	1000–30,000
EU	< 100	< 1000	< 10,000

Table 13 Types of water turbines used at hydroelectric power plants in Bulgaria

HPP	Turbine type	Impeller diameter (mm)
Momina Klisura	Francis turbines	3100
Orpheus		2250
Krichim		2000
Devin		2100
Kardzali		2520
Teshel		2000
Aleko		2000
Studen Kladenets		2300
Pasarel		860
Kokalyane		1440
Stara Zagora		1150
Zhrebchevo		1600
Rila		840
Vacha 1		1000
Vacha 2		860
Asenitsa		1000
Topolnitsa		1000
Pastra		740
Kitka		820
Tyzha		839
Lesichovo		650
Rositsa 2		840
Ivaylovgrad	Axial turbines	3700
Rositsa 1		2173
Koprinka		2173
Lakatnik		2850
Mezdra		1600
Sestrimo	Pelton turbines	2690
Belmeken		2090
Spanchevo		1960
Peshtera		1560
Popina Laka		1520
Pirin		1700
Lilyanovo		1800
Batak		2300
Beli Iskar		1430
Sandanski		1600

(continued)

Table 13 (continued)

HPP	Turbine type	Impeller diameter (mm)
Kalin		1100
Petrohan		1200
Kamenitsa		1300
Barzia		1500
Simeonovo		1055
Samoranovo		1250
Mala Zarkwa		900

hydrokinetic power plants, even of small groups of hydrokinetic turbines, as well as their management in real time, should be considered as a process of individual management of the turbines (rotors and each of their blades) aimed at maximizing the production of electric current as a whole.

4.1 Hydrodynamics of Turbines

The force interaction between a wing and the surrounding fluid is the basis of many engineering applications in aeronautical engineering, shipbuilding, turbomachinery and a number of others. Due to its importance, the matter has been studied very well theoretically and experimentally [17].

The contour obtained through a cross section of the wing is called the airfoil. In the simplest case, the wing can be uniquely defined by specifying the geometry of the airfoil and its length. In this case, the terms wing and airfoil are practically interchangeable.

When a fluid flows around an airfoil as shown in Fig. 13, a hydrodynamic force F arises, which can be decomposed into two components—lift force and drag force. These forces are represented by the following equations:

$$F_L = C_L b \rho \frac{w_\infty^2}{2} \tag{1}$$

$$F_D = C_D b \rho \frac{w_\infty^2}{2} \tag{2}$$

where C_L is the lift coefficient; C_D is the drag coefficient; ρ is density of the fluid; b is chord length representing the segment connecting the front and rear edges of the profile. The hydrodynamic characteristics of the airfoil are the dependences of the coefficients on the angle of attack.

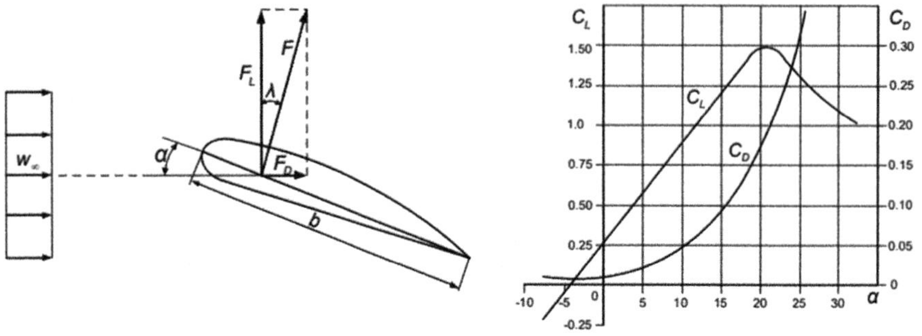

Fig. 13 Forces acting on an airfoil [17]

4.2 Reactive Torque

Apart from the mentioned two principles of the rotary movement of hydrokinetic turbines (hydrodynamic and resistance), there is also a third principle, which is used less often in turbines. This is the reactive principle of motion. The torque caused by it is in many cases commensurate with and even greater than the rotation, mainly due to hydrodynamic and drag forces.

With the help of Newton's third law, the formation of the reactive force is also explained. According to the third law of gas mechanics, a force equal in magnitude and opposite in direction opposes the rocket. Such a movement is called reactive, and the force that generates it called reactive, which is proportional to the flowing mass through the nozzle and its acceleration. The direction of the reactive force is opposite to the direction in which the fluid flows.

By properly directing the flowing water fluid for the hydrokinetic turbines, the direction of the reactive force can be also optimized [18]. The optimal direction is when it is directed orthogonally to the rotor radius, in the direction of rotation. The accelerated flowing water should be directed in exactly the opposite direction, behind the blades of the rotor blades. The acceleration is obtained thanks to the centrifugal forces that push the water from the center to the periphery of the rotors [19].

4.3 Test Rig for Low Pressure Turbines

In this part of the chapter, a test rig for low-pressure hydrokinetic turbines, which is located at the Technical University-Sofia, is described and presented [20]. The test-rig is shown at Fig. 14.

Figure 14 shows a test rig for testing low-pressure (kinetic) turbines and harnessing energy from rivers. The test rig allows testing kinetic turbines with impeller sizes from 50 to 320 mm. The rig comprises a technological and measuring part. The technological part includes the collection tank position 1. The volume of the collection tank is 2 m^3 and is consistent with the total amount of water needed to carry out the energy tests of the kinetic turbines. A single-stage centrifugal pump is installed to the collection tank, serving to create the desired parameters of the water in the measuring channel.

The flow parameters are adjusted with the help of the control valve 15. With the help of the mounted centrifugal pump we can adjust the velocity at the point required (the mounting point of the kinetic turbine) within the limits of 0.1–1.1 m/s. The diffuser 4 is installed in the initial part of the measuring channel, which acts as

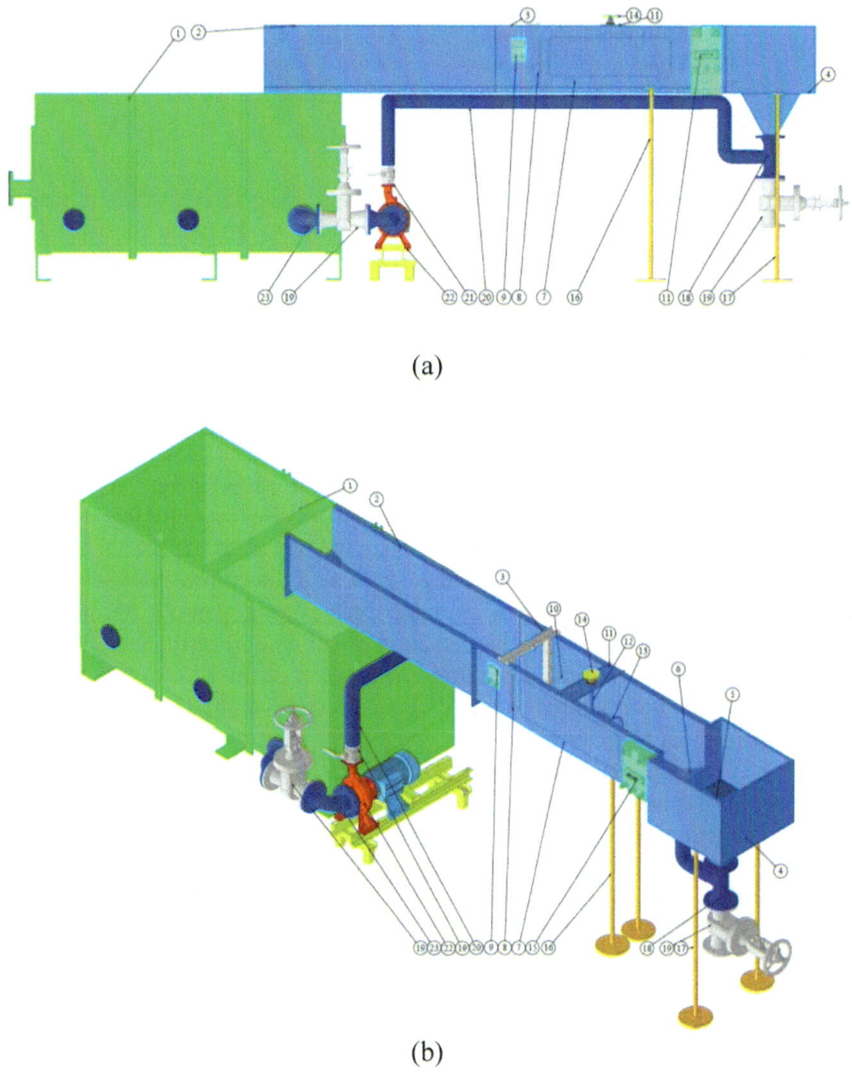

Fig. 14 Basic scheme of the test rig

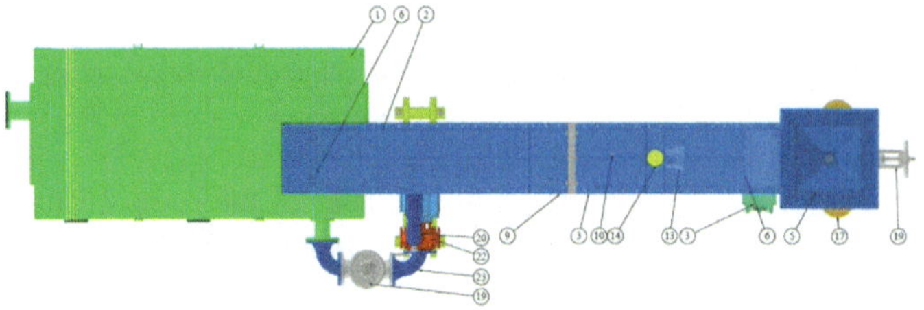

(c)

1 – Collecting tank; 2,3 - Channel segments; 4 – Diffuser;
5 – Stabilizing grid; 6 – Stabilizer; 7 – Observation hatch;
8 – Visual level gauge; 9 – Electronic level gauge with float; 10 – Turbine console;
12- Hydrokinetic turbine; 13 – Generator;14- Pump unit;
15- Control valve; 16 – Measuring board; 17- Inspection faucet.

Fig. 14 (continued)

the first flow stabilizer. After the diffuser, the water passes through the stilling grid 5 and the stilling 6. The channel where the energy tests take place consists of two segments, one of which is equipped with an observation hatch 7.

The overall dimensions of the channel are 3600/400/350 mm and they are adapted to the test velocities. After leaving the channel, the water returns back to the collection tank. The technological part also includes the console 10 for fixing the turbine and part of the measuring equipment. In the lower part of the diffuser, a revision valve is installed for emergency draining of the system.

The measuring part of the test rig includes a visual level gauge 8 for visually reporting the filling of the channel with water; Capacitive level meter (LW3B/SC10, range of measurement 0.5–4 m) with float 9 for more precise reading of the water level in the channel; Board for reading the data from the turbine generator 16 including measurement of rotation frequency, voltage, current, torque; Flowwatch tube (range of measurement 0.1–10 m/s), which is mounted on the turbine console. All measuring devices mounted on the test rig have their own calibration certificates. The tests carried out are within the time range of the calibration certificates.

The turbine unit is mounted on an axis connected to a reducer, which moves the shaft perpendicular to the turbine at an angle of 90° and transmits the rotation of the generator in the upper part of the channel. The gearbox that transmits the torque to the generator has a 1:1 gear ratio. During the turbine test, a diffuser was added to direct the flow to the working blades. Before placing the test turbine in the duct in its place, a velocity measuring device is installed exactly at the point of the turbine to measure the velocity of the flow at that exact point.

The methodology for conducting the tests is as follows: the pump unit is started and the flow parameters are set using the tap placed at the pump outlet. The parameters are set by continuously measuring the flow level using the digital level gauge and the velocity at the turbine insertion point using the velocity tube. The liquid level is regulated by placing plates with different useful cross section at the outlet of the channel. After reaching the desired velocity and flow level, the anemometer (flowwatch) tube is dismantled and the turbine is installed in its place.

The specially made bracket allows a stable mounting of the anemometer and the turbine unit so that the flow does not change their position during the measurement. After installing the turbine unit, the power supply coupling and the velocity sensor leading to the measuring board are attached to it. The frequency of rotation and the parameters of the generator are taken into account. The different loads (load) of the generator at a given rotation frequencies are achieved by including additional electrical consumers.

Experimental studies with three hydrokinetic turbines with different geometries, shown in Fig. 15 will be carried out on the test rig. The turbines will have the same diameter of 50 mm, but with a different number of blades and geometry of the blades themselves. Turbine (a) has 4 blades, turbine (b) consists of 12 blades, and turbine (c) has 6 blades. While the blades' profile of turbines (a) and (c) is the same, turbine (b) has different angle of attack, and larger number of blades with a smaller length. The active area of the blades is built according to their number applied for all three turbines.

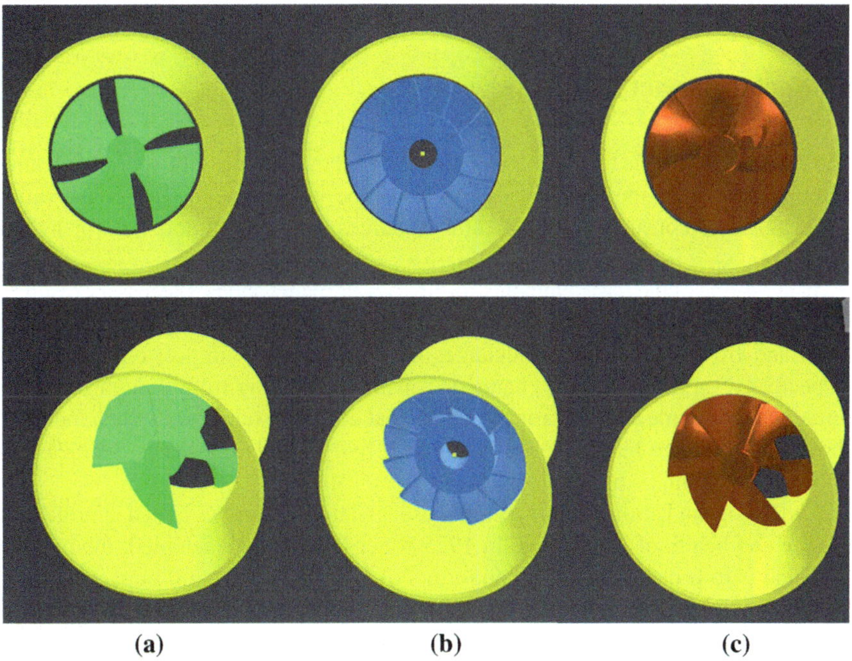

Fig. 15 Different hydrokinetic turbines' geometries

Table 14 Theoretical water turbine power

V (m/s)	P_t (kW)
0.2	0.03
0.4	0.05
0.6	0.07
0.8	0.18

The equation for determining the power is as follows:

$$P = \frac{C_D \rho A V^3}{2}, \text{kW} \tag{3}$$

where C is drag force coefficient (Fig. 13), ρ—density of water kg/m^3, A—clear section of the turbine, m^2, V—velocity, m/s.

In the studied case there are four different velocities, the drag coefficient is taken at an angle of attack of 25°, and the diameter of the turbine is 50 mm the four water velocities are actually the most commonly measured river water velocities in Bulgaria. Based on that, the theoretical turbine power is calculated. The results obtained are given in Table 14.

5 Social Impact

The role of a society is to prepare its citizens to behave responsibly not only towards other members of society, but also towards the environment. The conservation and reasonable consumption of natural resources is of key importance for human existence.

Water is one of the basic elements for life on our planet and is a necessary resource for the development of society and the economy, which need a sufficient amount of clean and fresh water. Climate changes have a significant impact on nature and, respectively, on man. Climate change is causing everything from droughts that can force a water regime and fail crops to floods that devastate our homes. Such incidents show the existing problems caused by human activity, but also the necessary measures to overcome them.

The European Union undertakes various legislative initiatives aimed at protecting the life and health of its citizens. In 1975 water legislation is adopted, the purpose of which is to protect consumers and users of water from harmful impacts. Access to quality drinking water is a key element of EU policy, which is reflected in various directives:

- Directive 98/83/EC, which has been revised by Directive (EU) 2020/2184, as of January 12, 2023 and aims to protect public health by monitoring the quality of water intended for human consumption.

- Bathing Water Quality Directive (76/160/EEC).
- Urban Wastewater Treatment Directive (91/271/EEC).

In 2015, Directive (EU) 2015/1787 was adopted, introducing new rules to improve the control of drinking water. The directive allows EU countries more flexibility in how they control drinking water across the Union. The Member States of the Union need to take the necessary measures to ensure that the water does not contain a concentration of micro-organisms, parasites or harmful substances that could pose a danger to human health and meets minimum microbiological and chemical standards. The changes in European water legislation are the result of the first European citizens' initiative, which runs the "Right to Water" (Right2Water) campaign. Thanks to greater transparency, more sustainable consumer choices are encouraged, such as drinking tap water instead of bottled water. Citizen initiative regarding the quality, protection and reasonable consumption of water as a resource is one of the prerequisites for facilitating the transition to a circular economy. Member States should make efforts to manage drinking water efficiently as a resource, which means using less energy and reducing water losses [21].

5.1 Water in Bulgaria

Water management in Bulgaria is carried out through the integration of international, European and Bulgarian legislation. International legislation includes all international conventions and agreements in the field of fresh water that a country has officially signed and ratified. There are three main international documents related to water: the Danube Convention, the Helsinki Convention, the Ramsar Convention on Wetlands and, in some aspects, the Black Sea Convention and Protocol for the protection of the Black Sea environment from pollution from land-based sources. In European legislation, the Water Framework Directive 2000/60/EU, which covers all other water directives, is of greatest importance for water management. The Water Framework Directive is the most important document for water management in Europe. In Bulgarian legislation, waters are protected by laws (Water Law) and regulations for their application. In Bulgaria, water resources are not abundant, which causes more and more serious problems. In the last year, water supply was disrupted in 17 regions, 40 municipalities, 12 cities, and 160 villages. Part of the reasons for the water regimes is related to the large losses along the pipelines, which amount to 500 million m^3 or about 56% [11]. In Europe, losses of about 20% are considered acceptable. Average water consumption per capita in Bulgaria was 102 L in 2020, which is higher than in the previous three years, when consumption was 99 L per day. The increased consumption is a consequence of the imposed quarantine due to Covid-19, which left people at home for a long period of time. Cities with more inhabitants also note higher daily water consumption, such as Sofia (127 L), Blagoevgrad (113 L), Burgas (108 L), Ruse (105 L), Plovdiv, Pazardzhik, Pernik (102 L) [11].

By comparison, in France this consumption exceeds 200 L per person per day, and Switzerland is the record holder with 300 L per day, 140 L are used in the United Kingdom [22]. The consumption of water resources for domestic needs is significantly lower compared to the water used in production. In our country in 2020 the most water resources are used in the "Industry" sector (3650.20 million cubic meters), followed by the water supply sector (366.12 million cubic meters) and agriculture (336.19 million cubic meters). Over 70% of the total amount of water used in Bulgaria in 2019 was intended only for cooling in the energy sector [11]. The data show that the largest consumer of water resources is industry, which requires a rethinking of production processes and a shift to more ecological productions.

5.2 Projects and Good Practices

Although water consumption for domestic purposes is significantly lower than that used in industry, it is extremely important that citizens become sensitive to the efficient use and management of water resources. Educating children regarding environmental and water protection is one of the main priorities of the EU and the member states in particular. A number of manuals and educational materials dedicated to environmental and water protection have been developed. The Ministry of the Environment and Water provides Bulgarian teachers with a guide to introduce young students to topics related to the protection of the environment and water. The materials are adapted to the age of the children and have a game form. The educational platform Ucha.se offers students animated clips that show the importance of natural resources for our lives and how important it is to be responsible for protecting water and the environment. The non-governmental sector in the country together with public and educational institutions conduct campaigns to promote responsible behaviour towards the environment and the reasonable consumption of natural resources. For example, ccampaigns such as "Clean environment", Policies for youth in the field of the environment, student competition "Ambassadors of health" of the Ministry of Education and Culture, "Old paper for a new book".

Greenwich and ECOPACK Bulgaria, "Let's clean up Bulgaria for one day" on BTV, etc., draw public attention to environmental problems and activate their behaviour. A positive trend is the activity of Bulgarian schools on the topic of water and environment protection. Introduced is "civic education" among students of all ages, looking at topics about the future of humanity and the need for responsible civic behaviour, promoting sustainable consumption and moderate consumption. In addition to the theoretical preparation, a number of activities on various international and local projects financed by EU programs (Erasmus+ program, ERDF) are also included. Good examples are: International project "Citizens—Civico Promoting civic competence among students" develops a handbook and materials for teachers in primary and secondary schools, through which to promote responsible behaviour among students. The "Water School" project aims to promote the conservation and sustainable use of water resources. Drinking tap water at school and reducing the use of plastic bottles is among the main activities of the project. The "Best water Use" project, in Greece-Bulgaria cooperation, aims to promote innovative technologies in order to improve environmental protection and the efficient use of water resources, as well as soil protection. International project "Water is life—let's protect it" aims to acquaint students with the sources of water pollution and ways of its rational use. Project "Saving water" aims to reduce the use of water resources in agriculture. It offers a training course "Saving water in agriculture".

The applied examples show a tendency to work with the young generation to develop knowledge and skills for sustainable consumption and committed behavior in water conservation and consumption. It is also necessary to promote civic awareness among the rest of the population. The use of energy and water consumption saving appliances; reducing the use of drinking water in agriculture, water consumption for personal purposes; reducing water losses from pipelines are only part of the measures that need to be taken. Citizens must become sensitive to the unscrupulous pollution of water bodies and their use in industry, because only in this way can sustainable development and conservation of resources be achieved for future generations [22].

6 Conclusions

The study presents how energy from running water can be used in the territory of the Republic of Bulgaria. The existing capacities built up to now are presented and their analysis is made. An experimental test-rig with its experimental equipment is presented and described in detail, which is used for the study of hydrokinetic turbines with the aim of harnessing the energy of the rivers of the territory of Bulgaria. This test-rig can be used to calculate the capacities of various geometries of hydrokinetic turbines. Based on them, as a future work, real turbines will be made, which will be tested in real conditions in different rivers on the territory of Bulgaria. The social effect of the use of river energy in Bulgaria is also briefly presented.

References

1. Ayadi F, Colak I, Garip I, Bulbul HI (2020) Targets of countries in renewable energy. In: 2020 9th international conference on renewable energy research and application (ICRERA). IEEE. http://doi.org/10.1007/s00170-021-06791-3
2. Naveen R, Revankar PP, Rajanna S (2020) Integration of renewable energy systems for optimal energy needs—a review. Int J Renew Energy Res 10:727–742. http://doi.org/10.20508/ijrer.v10i2.10571.g7944
3. Eurostat (2018) Production of electricity and derived heat by type of fuel. http://appsso.eurostat.ec.europa.eu/nui/show.do?dataset=nrg_bal_peh&lang=en
4. Eurostat (2020) Statistical Office of the European Communities, Luxembourg, SHARES 2018—short assessment of renewable energy sources. http://ec.europa.eu/eurostat/web/energy/data/shares

5. European Wind Energy Association (EWEA) (2020) Wind energy in Europe in 2019. WindEurope. https://windeurope.org/about-wind/statistics/european/wind-energy-in-europe-in-2019/
6. European Wind Energy Association (EWEA) (2020) Offshore wind in Europe—key trends and statistics 2018. https://windeurope.org/about-wind/statistics/offshore/european-offshore-wind-industry-key-trends-statistics-2019/
7. International Renewable Energy Agency (IRENA) (2019) Global energy transformation, a roadmap to 2050. www.irena.org/-/media/Files/IRENA/Agency/Publication/2019/Apr/IRENA_Global_Energy_Transformation_2019.pdf
8. International Energy Agency (IEA) (2020) World energy outlook 2019. Access to energy. https://www.iea.org/weo2018/themes/
9. REN21 Secretariat (2021) Renewables global status report. http://www.ren21.net/reports/global-status-report/
10. International Renewable Energy Agency (2022) Renewable capacity statistic 2022
11. National Statistical Institute (NSI) (2022) Water used by public water supply (WSS) households, average per capita—total for the country, by statistical regions, districts and basin districts
12. Hristova N (2012) River waters in Bulgaria. Paradigma, Sofia
13. National Institute of Meteorology and Hydrology (2022) Hydrological information and forecasts. https://hydro.bg/index.html. Accessed 18.12.2022
14. Lyubenov S (1990) Hydropower systems. Technika, Sofia
15. Obretenov V (2008) Water turbines. Ecoproogress, Sofia
16. Veiregistar (2022) https://www.veiregistar.bg/index
17. Dukov I (2020) Pumps, compressors and fans. Technical University of Sofia
18. Marinov M, Topalov I, Gieva E, Todorov G (2016) Air quality monitoring in urban environments. IEEE, Pilsen, Czech Republic, pp 443–448. http://doi.org/10.1109/ISSE.2016.7563237
19. Marinov MB, Topalov I, Ganev B, Gieva E, Galabov V (2019) UAVs based particulate matter pollution monitoring. IEEE. https://doi.org/10.1109/ET.2019.8878586
20. Velichkova R, Ivanov V, Stanilov A, Pushkarov M, Aleksandrov A, Simova I (2022) Modelling of a test-rig for the river energy utilization. IEEE. https://doi.org/10.1109/CIEES55704.2022.9990701
21. European Commission (2022) Environment. https://ec.europa.eu/environment/pubs/pdf/factsheets/water_scarcity.pdf
22. Smart Water Magazine (2020) Water ranking in Europe 2020. https://smartwatermagazine.com/news/locken/water-ranking-europe-2020

The manufacturer's authorised representative in the EU is Springer Nature Customer Service Centre GmbH, Europaplatz 3, 69115 Heidelberg, Germany. If you have any concerns regarding our products, please contact ProductSafety@springernature.com

Printed and bound by CPI Group (UK) Ltd, Croydon, CR0 4YY
01/10/2026
02225877-0011